Mechanics in
Structural Geology

BRIAN BAYLY

Mechanics in Structural Geology

With 230 Illustrations

SPRINGER-VERLAG

NEW YORK   BERLIN   HEIDELBERG   LONDON
PARIS   TOKYO   HONG KONG   BARCELONA   BUDAPEST

Brian Bayly
Department of Earth
  and Environmental Sciences
Rensselaer Polytechnic Institute
Troy, NY 12180-3590

Cover Art: Plate 59A from the *Atlas of Deformational and Metamorphic Rock Fabrics*, 1982 Springer-Verlag.

Library of Congress Cataloging-in-Publication Data
Bayly, M. Brian, 1929–
    Mechanics in structural geology / Brian Bayly. — 1st ed.
        p.   cm.
    Includes bibliographical references and index.

    ISBN-13: 978-0-387-97652-5   e-ISBN-13: 978-1-4613-9166-1
    DOI: 10.1007/978-1-4613-9166-1
    free paper)
        1. Geology, Structural.   2. Mechanics, Analytic.   I. Title.
    QE601.B36 1991
    551.8′01′531—dc20   91-17251

Printed on acid-free paper

Production coordinated by Chernow Editorial Services, Inc. and managed by Liz Corra.
Typeset by The Composing Room of Michigan, Inc., Grand Rapids, MI.

9 8 7 6 5 4 3 2 1

To W. B. H.

Preface

People respond to the natural world in two ways. When an extensive view unfolds ahead of us, we feel good. Or when we see a natural arch or a beach stone with a hole through it, we cogitate, wondering: How did that happen?—we feel and we ponder.

These two responses developed over different time scales. Organisms have been reacting to their environment for at least three billion years. Times have often been hard, and only those individuals with special traits survived. But enjoying the world around is a survival trait, so that the present population contains mainly people who enjoy being here. The feeling that the natural world is good and beautiful, a pleasure to look at, has been strengthened in us by breeding over this very long time.

The pleasure of pondering is a more recent development. For perhaps three million years, people have been enjoying, from time to time, that moment of elation when we first notice a connection: Hey, I never realized until this moment how this links up with that!

These two pleasures—one old and probably common among animals, one newer and probably confined to man—are the two generators for this book; I would like to share them with other people. I have myself been fortunate; people led me among mountains and explained what there was to see; the sunlit textures looked good, and occasional glimpses of reasons felt good. Mechanics in structural geology is not academic or dry, it is an approach to a throng of marvels to see and puzzles to ponder, as rich as any within the scope of our natural senses. I hope that readers get increased pleasure from looking at rocks, and that they in turn will open new vistas for me to enjoy.

A word is in order about the organization of the material. Chapter 2 concerns geometrical changes, whereas Chapters 3 and 4 are about the forces and stresses that produce them; Chapters 5, 6, and 7 are about material behavior—how it is that in some materials a small stress produces a large geometrical change, while in other materials a larger stress produces a smaller change. The chapters should be read in sequence except that the sequence 3, 4, 2 can be followed if a reader prefers it.

Readers are encouraged to be selective; for example, it is not necessary to read the whole of Chapter 2 before going on to Chapter 3. If one thinks of Chapter 7 as the most entertaining, one should pick from Chapters 2, 3, and 4 just enough to understand 5 and 6, and enough from 5 and 6 to understand Chapter 7. No single scheme can serve the diverse interests of all readers but the diagram on the inside front cover of this book gives a partial indication of the links between sections.

Regarding the questions that are dispersed through the text, it is recommended that readers work on these as they come up, answering the question and then comparing with the answer provided before going any further with the text. Some points that need to be understood are introduced by means of a question and its printed answer; readers who skip the question-and-answer loop may find the text that follows hard to understand, because they miss learning some necessary concept. Structural geology is not a spectator sport. This book is designed for people who want to be active participants; there should always be a pencil and scratch pad handy.

Acknowledgments

Every writer is indebted to other writers; I have learned much, and have much more still to learn, from books and papers by fellow workers.

People who used early versions of the text have been an inestimable help. In particular, people who were students in my classes during the gestation years were the stimulus for, as well as the cheerful critics of, these pages. Along with that of Frank Florence, their guide and consultant, I wish their contribution to be recognized to its full extent. I hope only that the product is worthy of the contributions they made.

Four people in particular have opened their minds to me—kind accidents of fortune have let me learn from them: Brian Harland, John Jaeger, Hans Ramberg, and Win Means. When I reflect on how the flame of scholarship is passed from one generation to the next, these are my exemplars; it is a pleasure to acknowledge my debt.

Contents

Chapter 4

Variation of Stress with Direction 89

Chapter 5

Rheology: Relations Between Forces and a Material's Response *111*

Mechanics in
Structural Geology

CHAPTER 1

Introduction

The topic *mechanics in structural geology* has an impersonal ring, but this book is intended for *people*. It springs from the belief that people are naturally curious about what is around them; just one pinpoint and example within the range of human curiosity is the question where mountains come from.

MOUNTAINS AND HUMAN AFFAIRS

The frame for all human activity is the earth's surface. Two streams of energy interact to mold that surface—the solar flux and the flux of earth's internal heat. The first is the more vigorous, by a factor of about 7000. Nonetheless there is one respect in which the two fluxes roughly balance each other: the internal flux builds mountains up at roughly the same rate that sun-driven processes wear them down. The balance is of course only approximate. At any moment there are variations from place to place, at any locality there is variation from time to time, and for all we know there may be an overall trend toward a smoother, less rugged earth. But the average roughness of the earth's surface does not change very fast; mountains have probably been present throughout the evolution of human life, influencing the climate and the occurrence of forests and fertile plains, our habitat.

Besides providing our habitat, mountains provide a window into the earth's crust. Desert sand or a fertile plain conceals the nature and behavior of the crust, whereas mountain belts expose the crust to view. The processes that smooth the earth's surface can be studied anywhere on that surface, but the processes that roughen it are best studied in present mountains or the eroded roots of their precursors. There we find the *structures* of structural geology, and gather clues to the processes and driving forces.

Not to take too narrow a view, structures in submarine sediments, both present-day and those preserved from former times, are a second source of clues to how rocks behave. But present-day submarine sediments require oceanographic study methods; it is their lithified equivalents that provide clues more readily. The clues are found in deformed materials in terrestrial outcrops, and it is here that the structural geologist comes to grips with them. Deformed structures in present-day mountains or old eroded remains help us to learn about our habitat.

MOUNTAINS AND THE CONCEPTS OF MECHANICS

The observables and the inquiring geologist somewhat resemble a block and an intending sculptor: the natural condition is almost formless, it is by human activity that an intelligible form is made to emerge. In the same way that a sculptor parts the marble with her tools, so the structural geologist works on the mass of observables; a probing mind separates one aspect from another, like a sculptor's blade.

Distinguishing the sculptor's tools from the block on which she works brings out a characteristic of this book: our prime concern is with the tools rather than with the material. Many fine accounts exist of the structures that exist in mountains, but there are fewer surveys of the intellectual equipment one might use when working on them. The piece of equipment of particular concern is familiarity with the concepts of mechanics. If this equipment is in one's mental toolbox, some tasks can be accomplished that cannot be done without it, or can only be done less well: switching from sculpture to woodworking, one can imagine a toolbox that contains axes but no saw.

Naturally, illustrative examples are used as well as pieces of theory, and the preface contains suggestions about how to use the various components of the book. Separately from those suggestions, the purpose of this introduction is to show the direction in which we are headed or the state we hope to attain, and familiarity is the key. The equations of mechanics as old friends, or as tools that fit easily in the hand from previous use—this is how we wish to possess them. Everybody knows a certain amount of mechanics: the objective is to make more ideas familiar and to increase the pleasure that comes from practicing their use.

Strains and Displacements

Geology is very much concerned with spatial arrangements: the movement of Eurasia with respect to America, the location and extent of an oil field or ore deposit, the ups and downs of topography and their influence on drainage, groundwater and so on. And of course the interesting parts are not when everything is standing still; it is *change* of position, rearrangement, and *change* of shape that catch our attention and cry out to be understood. Our primary concern in this chapter is change of shape.

Figure 2.1 shows an intuitive separation of three ideas—change of position, change of orientation, and change of shape. Of course in diagram (c), individual points change position: the essence is that in (c) they change position *with respect to each other,* whereas in (a) and (b) the change is only with respect to some external reference frame. It is diagram (c) that we focus on, where the parts move with respect to each other. In such a change, several things go on simultaneously and our objective is to get a crisp statement of these things in numbers. As mentioned in the Introduction, it is exchanges of energy that control earth processes and our lives, and a numerical description of change of shape is a start toward assessing what change of energy might be involved.

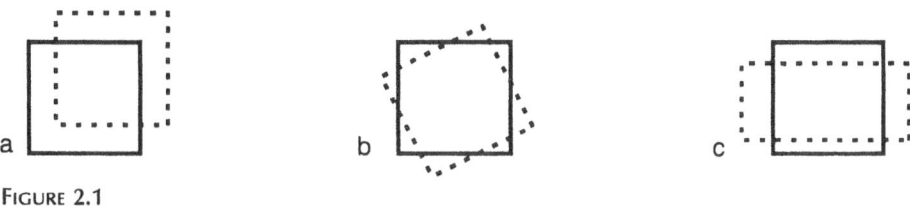

FIGURE 2.1

LINEAR STRAIN

If two particles in a continuous piece of material, such as two grains in a slice of bread, are 5 cm apart at one moment and 6 cm apart later on, one says that the strain is 1/5

$$\textbf{linear strain} = \text{(change of length)/(initial length)}. \qquad (2.1)$$

A change from 5 cm to 4 cm would also involve a strain of 1/5, and the convention adopted here is to call **elongations** positive strains and **shortenings** negative strains. Because it is the ratio of one length to another length, strain is dimensionless, just a number.

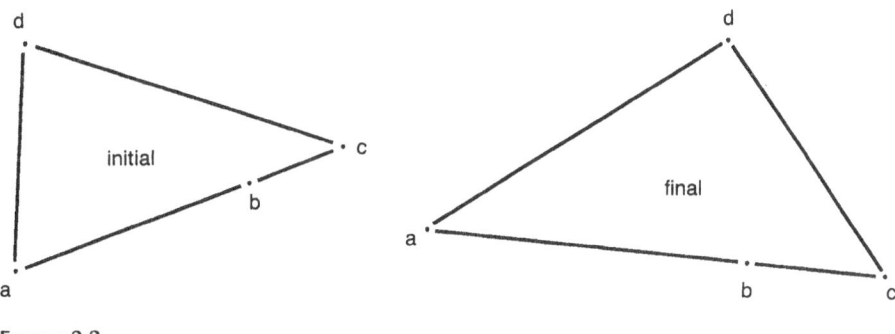

FIGURE 2.2

Question 2.1 Linear strain: Estimate the linear strain for *ab, bc, ac,* and *dc* in Figure 2.2.

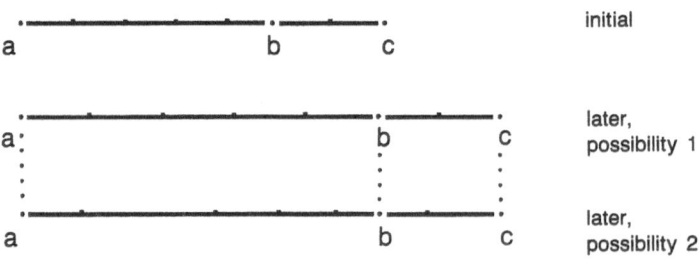

FIGURE 2.3

An outcome of Answer 2.1 is that the strain along line *ac* is not uniform; see Figure 2.3. One possibility is that, even though the strain is not uniform over the whole length *ac*, it is uniform over the separate sections *ab* and *bc*. On the other hand, as shown in possibility 2, the strain might vary considerably all along the line. In this last situation, the thing we can do with greatest confidence is to look at a whole series of very short portions of the total line *ac*. The shorter the portion examined, the less the strain can vary within it, and a new definition,

$$\text{linear strain} = \text{limit of} \left\{ \frac{\text{change of length}}{\text{initial length}} \right\} \text{ as initial length approaches zero,} \quad (2.2)$$

takes away any uncertainty. Equation (2.1) gives the **average strain** over an interval. Equation (2.2) gives the **local strain** at a point. In an important special case, the local strain is the same at all points within an interval and thus the local value and the average value are the same, but this ideal condition is rarely met in geology. We often estimate an average strain, over a few millimeters or kilometers, but it is only rarely that we go on and look at the local strain magnitudes that contribute to the average. The converse is more often done: we gather local estimates and compile them into a more regional average.

Question 2.2 Beds made thinner by solution: The following stratigraphic column shows the original thickness of four formations, and the amount by which each formation has been thinned by solution processes. What is the average thinning ratio for the whole column?	*Original thickness*	*Fraction removed by solution*
	80 m	0.10
	48 m	0.15
	10 m	0.66
	40 m	0.02

Question 2.3 Alternate shale and limestone: A sedimentary succession was originally 2/5 shale and 3/5 limestone, and has lost by dissolution 0.12 of the shale and 0.18 of the limestone. What is the average or overall loss-fraction?

The result from Answer 2.3 is an example of a **weighted mean.** It is a mean of the two separate loss-fractions, 0.12 and 0.18, but it is weighted in favor of the more abundant constituent (limestone), and hence comes out closer to 0.18 than to 0.12. Any time that local strain values are combined into a regional average, they need to be weighted in this manner to allow for the extent of rock that each local value applies to.

Summation and Integral Forms The ideas just introduced reappear in other contexts, and compact expressions are a convenience. The idea in Answer 2.2 can be expressed as follows. Consider n layers, each identified by an integer i, with i taking values from 1 to n. For the ith layer, let the thickness be t_i m and the loss-fraction be α_i. Then the thickness lost from layer $i = \alpha_i t_i$ m, and the total loss from the stack of n layers, measured in m, is $\Sigma_{i=1}^n \alpha_i t_i$. Also, the original total thickness $= \Sigma_{i=1}^n t_i$ m, so that the overall loss-fraction equals $\Sigma \alpha_i t_i / \Sigma t_i$. This manner of writing a weighted mean is sufficiently common that it is as well to be familiar with it.

A variant on the preceding calculation is as follows. Let the total thickness $\Sigma_{i=1}^n t_i$ be designated T. Then for any single layer, a thickness-fraction exists, equal to t_i/T, the fraction of the whole stack that that layer occupies. Call this f_i; then the overall loss-fraction is $\Sigma_{i=1}^n \alpha_i f_i$. Let us check this immediately against Question 2.3: Values of α and f for the shale are 0.12 and 2/5, and for the limestone 0.18 and 3/5. The summation is $(0.12)(2/5) + (0.18)(3/5) = 0.156$, in agreement with the answer reached by directly imagining a real outcrop.

In Figure 2.3, possibility 2 suggests that the strain varies in a continuous manner all along the line, rather than being uniform in each of n segments of finite length. The total change of length would then be $\int_0^L e \cdot dl$, where L is the total length of line, and e is the strain in a segment of infinitesimal length dl at distance l from one end. To keep track of the fact that in this expression e varies with l, it is often helpful to write $\int_0^L e(l) \cdot dl$. Corresponding expressions for the overall or average strain would be

$$\frac{\int_0^L e(l) \cdot dl}{L} \quad \text{or} \quad \frac{\int_0^L e(l) \cdot dl}{\int_0^L dl} \quad \text{or} \quad \int_0^L e(l) \cdot dl/L.$$

Compare these with

$$\frac{\sum^n \alpha_i t_i}{T}, \quad \frac{\sum^n \alpha_i t_i}{\sum^n t_i}, \quad \sum^n \alpha_i (t_{i/T}) \quad \text{or} \quad \sum^n \alpha_i f_i.$$

Question 2.4 Summations:

(a) Suppose one invests n sums of money s_i ($i = 1,2,\ldots,n$) each at a rate of interest r_i per year. What is the overall rate of return on the set of investments?

(b) If a volume V of rock is made up of a set of infinitesimal elements dV each with density $\gamma(v)$,

what is the total mass of rock? what is its average density?

(c) If a rock contains n minerals each constituting a fraction f_i of the whole volume of rock ($i = 1,2,\ldots,n$) and each mineral's density is γ_i, what is the density of the rock?

Strain Rate If a small strain Δe occurs in a small time Δt, the strain rate is $\Delta e / \Delta t$. The small strain in turn is $\Delta l / l$ where l is the length of some line in the material whose behavior is being described. In the infinitesimal limit,

$$\text{strain rate} = \frac{de}{dt} = \frac{dl}{l \cdot dt}. \tag{2.3}$$

Newton's dot notation is often used for strain rates; that is, $\dot{e} = de/dt$. Let us note that the idea of strain involves a definite initial state and final state, but the idea of strain rate does not involve identifying any particular initial state; the length l in Equation (2.3) is the *current* length of the line being described.

Question 2.5 Stretching gum bar: The table shows the history of elongation of a bar of gum:

After time	0	3	7	8	9	10	11	12	min,
Length was	5.0	5.9	8.1	8.7	10.2	11.6	13.5	broken	cm.

For the eighth minute, we have "initial" length = 8.1 cm, final length = 8.7 cm, strain = 0.6/8.1 = 0.074 and so the average strain rate during that minute was 0.074 per minute. Compare this value with the strain rate at the moment the eighth minute began, when the length was 8.1 cm.

Suggestion: Plot length against time and sketch a continuous smooth line through the data points; the gradient of such a line has a recognizable value *at any instant*. But the gradient is not exactly the quantity we seek: the gradient has units centimeters per minute (cm/min), whereas the strain rate is a plain number per minute.

Question 2.6 Constant strain rate and varying strain rate:

(a) If a line gets longer in such a way that its strain rate is constant, how does its length change with time? (That is, how would a plot of length versus time look?)

(b) If a line gets longer at a steady rate in centimeters per minute, how does its strain rate change with time?

Strain rate resembles compound interest. In either case, if the change over one period of time is by a factor of 1.03, for example, the change over n periods of time is $(1.03)^n$.

Logarithmic Strain   We just noted that if strain rate remains constant through n periods of time, the total factor of change is $(1 + \dot{e})^n$. Call this F; then $\log F = n \cdot \log (1 + \dot{e})$. If $\dot{e}$ is small in comparison with l and we use natural logarithms, we can use the approximation $\log_n (1 + \dot{e}) = \dot{e}$ and conclude that

$$\log_n F = n \cdot \dot{e}. \tag{2.4}$$

Thus the quantity gained by multiplying the strain rate by time (e.g., 0.03 per year multiplied by 20 years = 0.6) is $\log_n$ (final length/initial length).

Question 2.7 Cumulative strain: The bar in Question 2.5 elongated during the eighth minute at about 0.075 per minute or 4.5 per hour or 0.00125 per second. If you estimate the total change during a 5-minute period at this *steady* strain rate, does it matter which expression of the strain rate you use?

As can be checked, the bar's strain rate did *not* remain steady through time, and there is no reason to suppose that strain rates in rocks ever remain steady either. Equation (2.4) is in fact something of a trap; the assumption it embodies is almost always not a very strong assumption.

Stretch   A quantity closely related to a line's strain is its stretch:

$$\textbf{stretch} = (\text{new length})/(\text{initial length}) = l + \text{strain}. \tag{2.5}$$

As the previous section showed, certain geometrical truths are more conveniently expressed in terms of stretch than in terms of strain; but a material's mechanical behavior—its response to an imposed stress—is usually expressed in terms of strain, and consequently the latter concept is the more extensively used in later chapters.

Question 2.8 Regional stretch: In a mining region, parallel planar pegmatite veins recur about five every kilometer, and they average 1.8 m wide. By what factor has the region stretched in a direction normal to the veins?

Change of Area and Volume

Question 2.9 Area strain: A rectangle measuring 5 m north-south by 7 m east-west suffers a north-south elongation of 0.12 and east-west shortening of 0.08 (i.e., strain = −0.08).

(a) What is its new area?

(b) What is the change of area?

(c) What is the area strain, if **area strain** is defined as (change of area)/(initial area)?

(d) Answer part (c) again for a rectangle measuring Jm north-south by Km east-west.

Is there a direct numerical relation between Answer 2.9 (c) and the two given strains, 0.12 and −0.08? There must be a relation, but it is more directly seen if we work with the stretches north-south and east-west, which are 1.12 and 0.92, respectively. Their product is 1.0304 = (new area)/(initial area) and in fact rather obviously in general

$$\frac{\text{new area}}{\text{init area}} = \frac{\text{new length}}{\text{init length}} \times \frac{\text{new breadth}}{\text{init breadth}} = \frac{\text{stretch}}{\text{north-south}} \times \frac{\text{stretch}}{\text{east-west}}$$

If we call the left-hand term the **area stretch** S_A, then

$$S_A = S_{NS} \cdot S_{EW}, \tag{2.6}$$

a very simple and compact result. The corresponding statement in terms of strains is clumsier but somewhat instructive, as follows:

$$\text{new dimension north-south} = J(1 + e_{NS});$$

$$\text{new dimension east-west} = K(1 + e_{EW});$$

$$\text{new area} = J(1 + e_{NS}) \cdot K(1 + e_{EW});$$

$$\text{change of area} = J(1 + e_{NS}) \cdot K(1 + e_{EW}) - \text{initial area } JK;$$

$$\text{area strain} = \frac{\text{change of area}}{\text{initial area}} = (1 + e_{NS})(1 + e_{EW}) - 1$$

$$= e_{NS} + e_{EW} + (e_{NS} \cdot e_{EW}) \tag{2.7}$$

(Check this result by inserting the values from Question 2.9: area strain = 0.12 + (−0.08) + (0.12)(−0.08) = 0.0304, as before.) The main value of this result is in the special case when the strains are small—1 or 2 percent or less; in such a case, the product of two strains can be neglected and the area strain = $e_{NS} + e_{EW}$. To sum up:

---

area stretch = $S_{NS} \cdot S_{EW}$ exactly, even if stretch is large;   (2.6)

area strain = $e_{NS} + e_{EW}$ approximately, and only when strains are small. (2.8)

Rates of Change   As was already remarked, a material's response to an imposed stress is often given in terms of strain or strain rate. A **strain rate,** the rate at which the material is straining at some moment, necessarily involves the idea of a *small* strain. The linear strain rate is in fact the limit of the ratio $\Delta e / \Delta t$ as Δe and Δt both approach zero, where Δt is a small interval of time and Δe is a small linear strain occurring in that time, $\Delta l / l$ (current length). Because the concept involves going to the limit where the strain is small, for *rates* of strain the summing process is exact:

$$\text{area strain rate} = (\text{strain rate})_{NS} + (\text{strain rate})_{EW}$$
$$dA / A \cdot dt \tag{2.9}$$

exactly, but only at the instant of observation.

Conservation of Area and Volume   The results that have been given for change of area can be extended to change of volume. We use 1, 2, and 3 to designate three orthogonal directions (instead of height, length, and breadth). Then:

$$
\begin{aligned}
\text{volume stretch} &= S_1 \cdot S_2 \cdot S_3 \text{ exactly,} \\
\text{(new vol)/(init vol)} &\quad \text{even if stretch is large;} \tag{2.10}
\end{aligned}
$$

$$
\begin{aligned}
\text{volume strain} &= e_1 + e_2 + e_3 \text{ approximately,} \\
\text{(change of vol)/(init vol)} &\quad \text{only when strains are small;} \tag{2.11}
\end{aligned}
$$

$$
\begin{aligned}
\text{volume strain rate} &= \text{strain rates } \dot{e}_1 + \dot{e}_2 + \dot{e}_3 \\
dV / V \cdot dt &\quad \text{exactly, but only at the} \\
&\quad \text{instant of observation.} \tag{2.12}
\end{aligned}
$$

These equations are particularly simple to apply to those circumstances where we can assume that the net volume of material is not changing. Many geological processes are simply rearrangements of grains or rock masses (or at least can be treated as such), so that the left-hand side is precisely 1, in (2.10), or 0, in (2.11) and (2.12).

Question 2.10 Deformations that conserve volume:

(a) If two orthogonal stretches are 1.35 and 0.64, and volume is conserved, what is the third stretch?

(b) If two orthogonal strains are 0.003 and −0.002, and volume is conserved, what is the third strain?

(c) If two orthogonal strain rates are −0.01 per year and 0.05 per year and volume is conserved, what is the third strain rate?

(d) Repeat part (c) for strain rates −10 per 1000 years and 50 per 1000 years. What is the third strain rate?

(e) Try drawing a cube and a brick in perspective to illustrate your answer to (a) and see if it "looks right."

An even more special case, but nonetheless one that is much discussed, is **plane strain,** where a material (clay, rock, etc.) is shortened in one direction and elongated in a perpendicular direction, while the third orthogonal dimension does not change at all ($e_3 = 0$); a fistful of uncooked spaghetti is readily treated in this manner. If an initial rectangle changes to a different rectangle—as suggested in Figure 2.1(c), our prototype for change of shape—then at constant volume and constant dimension perpendicular to the plane observed, the change must conserve area.

In plane strain at constant volume, area is conserved and

$$S_1 \cdot S_2 = 1, \tag{2.13}$$

$$e_1 + e_2 \cong 0 \quad \text{for small strains}, \tag{2.14}$$

$$\dot{e}_1 + \dot{e}_2 = 0 \quad \text{at any instant}. \tag{2.15}$$

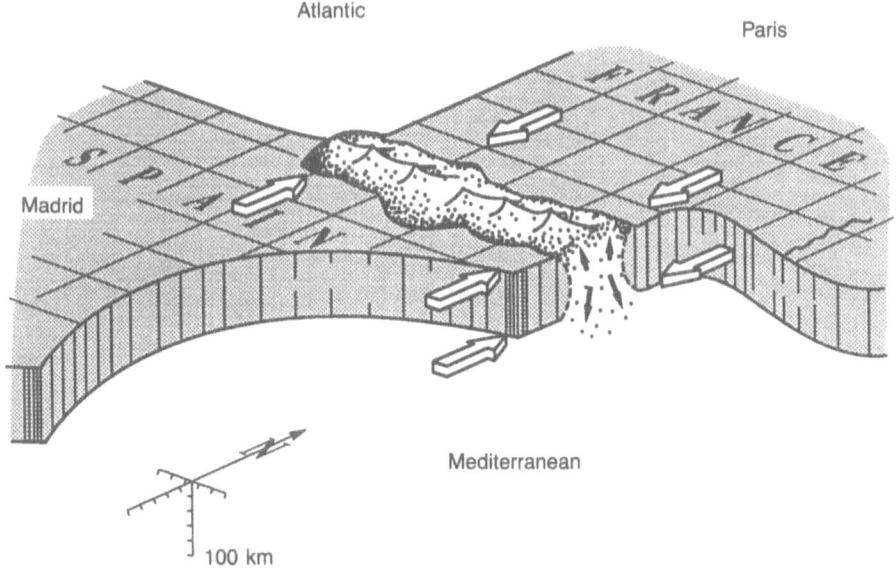

FIGURE 2.4 Formation of the Pyrenees. Since this figure was drawn, geophysical studies have revealed noticeable asymmetry in the deep structure of the Pyrenees, but a symmetrical picture continues to serve best for introducing the ideas we are concerned with in this text.

Question 2.11 The Pyrenees: It has been suggested that in the formation of the Pyrenees, Spain and France moved 90 km toward each other during a period of about 10 million years (See Figure 2.4). If the motion were steady, this would be a relative displacement of 9 mm/yr.

(a) Convert this to a linear shortening rate, at a time when the "rigid blocks," France and Spain, were separated by a 180-km-wide strip of deformable material. (Give the rate per year, per million years, and per second.)

(b) Your answer to (a) is a hypothetical average over the whole deformable strip, but assume that it is also a true local value for some portion of the strip. Assume also that volume is conserved and that the length of the Pyrenees from the Atlantic to the Mediterranean did not change. What is the local strain rate in a vertical direction?

(c) If the continental crust in the location considered is 40 km thick, at what rate in millimeters per year is the top of the crust moving upward with respect to the base?

Spread Ratio   In addition to plane strain, another special situation worth noting is one with cylindrical symmetry: all directions perpendicular to some line in the material strain at the same rate (see Figure 2.5).

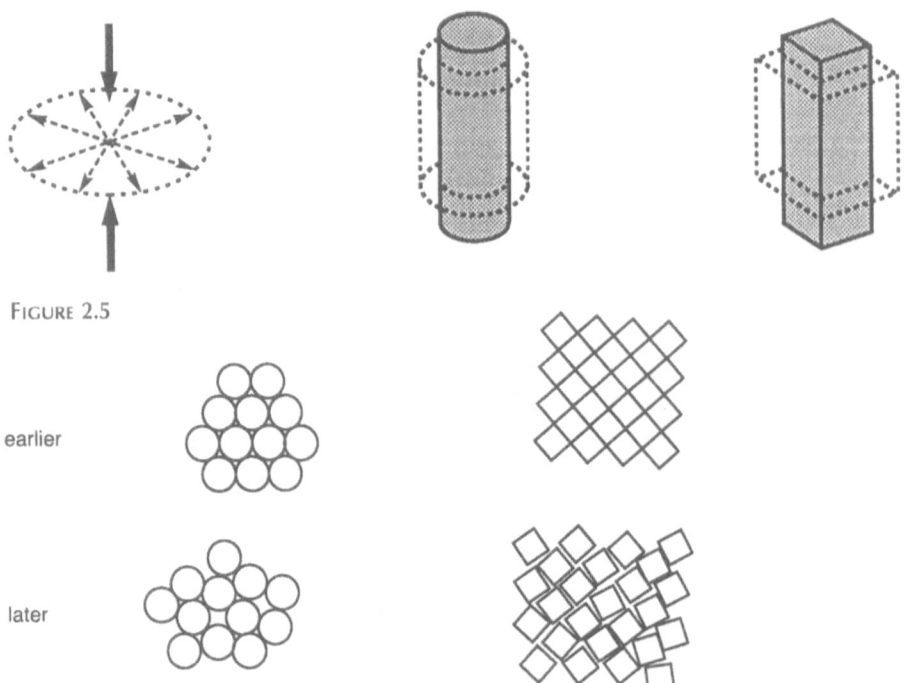

FIGURE 2.5

earlier

later

FIGURE 2.6

For the purpose of using the relations already noted, any pair of orthogonal lines in the plane of spreading can be used as reference directions 2 and 3; then $e_2 = e_3$ and to conserve volume we need $e_2 = e_3 = -(1/2)e_1$. But volume is not always conserved. Compacting sediments usually do not spread horizontally while their vertical dimension diminishes; i.e., $e_2 = e_3 = 0$. The values 0 and 1/2 are in fact useful limits: most natural instances, though not all, fall within this range. Synthetic foams like styrofoam or bread have spread ratios close to 0; so also do natural foams like pumice. "Simple fluids" like honey and glacier ice have ratios close to 1/2, and some cheese has a ratio less than 1/2. An important exception to the limits given is a class of materials with ratio greater than 1/2, **dilatant** materials, which take up a *larger* volume when deformed (see Figure 2.6). It is fortunate that most beach sand is dilatant, with ratio > 1/2, since water-filled sand with ratio < 1/2 is quicksand. Dilatant behavior of fractured rock in fault zones is an important influence on fluid effects in earthquakes.

Readers accustomed to Poisson's ratio will realize that it is a special case of the spread ratio discussed here, for materials and deformations where the strains totally reverse themselves on release of stress. In geology, constriction is also fairly common ($e_2 = e_3$ = shortening), and an extrusion ratio with usual limits 0 and 2 could be used.

Question 2.12 Spread ratio: If the strain that drives spreading (the axial strain) is a shortening of -0.04 and the spread ratio is 0.3, what is the volume strain? If we invent a factor F so that (volume strain) $= F \times$ (axial strain), what is F when spread ratio is 0.3?

Question 2.13 Extrusion ratio: In a constriction process, the radial shortening rate is 0.2 per million years and the extrusion ratio is 1.6. What are the elongation rate and the rate of volume strain?

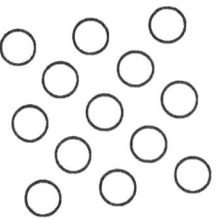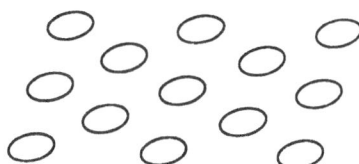

FIGURE 2.7

Variation of Linear Strain with Direction

Up to this point we have given attention either to a single direction in space or to two or three orthogonal directions. Orthogonal lines have the special property that change of length of one is wholly independent of change of length of another, but we have only to think of a triangle to realize that this is not the general case. In general, given several lines in a material, their strains are related to each other—and this is the next topic to study. But as a preliminary we need the idea of a homogeneous strain.

Homogeneous Strain To define uniform strain along a line, we imagined many small divisions and stated that for uniform strain, the strain in each division must be the same. Similarly with an area or a volume, if it can be divided into many divisions and the strain is the same in each, the strain is said to be **homogeneous** within the area or volume; see Figure 2.7.

Useful associated ideas are that in homogeneous strain

1. straight lines remain straight;
2. parallel lines remain parallel;
3. circles become ellipses and spheres become ellipsoids. Also, a set of uniform circles becomes a set of uniform ellipses.

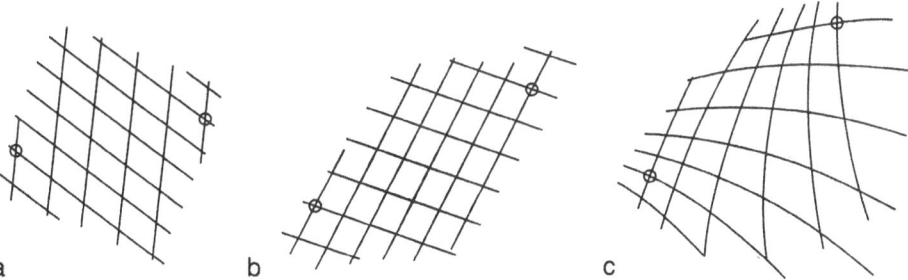

FIGURE 2.8

Question 2.14 Homogeneous and inhomogeneous strain: Use Figure 2.8. Diagram (a) is a grid with no special properties of spacing or orientation (except uniformity) and diagrams (b) and (c) are deformed versions of the same grid. Draw a circle on grid (a); for every point where the circle intersects a grid line, mark a corresponding point on grid (b) and grid (c), and so construct a continuous loop on each grid that is the deformed version of the original circle.

The loop on grid (b) should be an ellipse. Measure its long axis and its short axis and the diameter of the circle you drew on grid (a). Hence calculate a stretch and a strain for the long axis, and a stretch and a strain for the short axis.

Principal Axes   The operation in Question 2.14 brings out a fact on which the whole of continuum mechanics depends. The two grids, Figure 2.8(a) and (b), are in no way special; any time that one uniform grid becomes another uniform grid, we have homogeneous deformation, and two numbers specify it completely (or in three dimensions, a sphere would become an ellipsoid and three numbers would arise).

Question 2.15 Representation of strains by ellipses: Start with a circle either 1 in. in diameter or 2 cm in diameter. Draw four ellipses that would result from imposing on the initial circle the following changes:

    (a) strains of +0.4 and +0.6;
    (b) strains of +0.5 and −0.3;
    (c) stretches of 1.5 and 0.9;
    (d) stretches of 0.8 and 0.6.

The circle and ellipse at Question 2.15(a), for example, are a total representation of the change of shape *and* the two numbers are a total representation also. In a homogeneous deformation, the only shape a circle can turn to is an ellipse and, to specify an ellipse, two axial lengths are all one needs, since the axial directions have to be perpendicular to each other. The axial directions are known as the **principal axes** of the deformation.

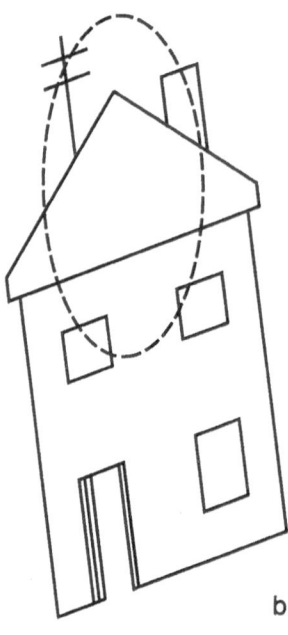

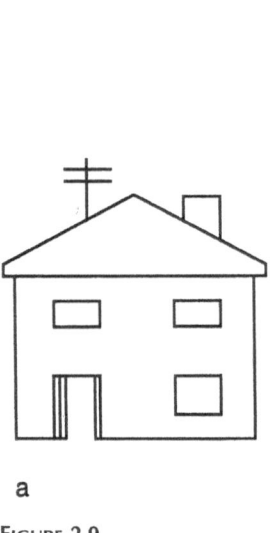

a   b

FIGURE 2.9

Question 2.16 Principal axes: What is the angle between the principal axes at the start of the deformation?

(a) Refer to Figure 2.8 again. The principal axes are visible in any ellipse you have drawn in grid (b); by locating corresponding grid intersections, insert lines on grid (a) that correspond to the ellipse axes in grid (b) and measure the angle between them.

(b) Refer to Figure 2.9. Mark the principal axes of the ellipse in part (b) and use the details of the scene to locate the original position of the corresponding lines in part (a); measure the angle between them.

In concluding this section, it is to be recalled from the start of the chapter that we are not using any external reference coordinates; we are discussing here movement of material particles in a rock with respect to each other. To reinforce this point, we restate what was discovered at Answers 2.14 through 2.16:

> In a **homogeneous deformation,** the line of material particles that suffers greatest strain and the line of particles that suffers least strain make a 90° angle at the end of the deformation, and also make a 90° angle at the start of the deformation.

It is highly unusual that these two lines of particles maintain a 90° relation all through the deformation. In fact, it is just an idealization to suppose that they ever do. However, that idealized process is often discussed and is called a **coaxial deformation.**

Lines Inclined to the Axes   Figure 2.10 shows an initial circle, an ellipse that is produced from the circle by a homogeneous deformation, and a series of material lines *Oa, Ob, ...,* etc. that are uniformly spaced in the original circle. Each of these reappears in the deformed state where its linear strain can be estimated. *Od* is the line of least strain (i.e., least elongation or maximum shortening, with the smallest numerical value of strain) and *Om* is the line of greatest strain. As we consider in turn lines *Od, Oe, ...,* etc., we must necessarily find amounts of linear strain that increase from the minimum, *Od,* up toward the maximum *Om* and then begin to decline again through *On,* etc.; that is to say, the behavior must be somewhat as shown in Figure 2.10(c). Is this profile a sine wave? For large strains, it is not; the smaller the strains, the more closely the profile approximates a sine wave, and for strain rates, the curve is exactly a sine curve.

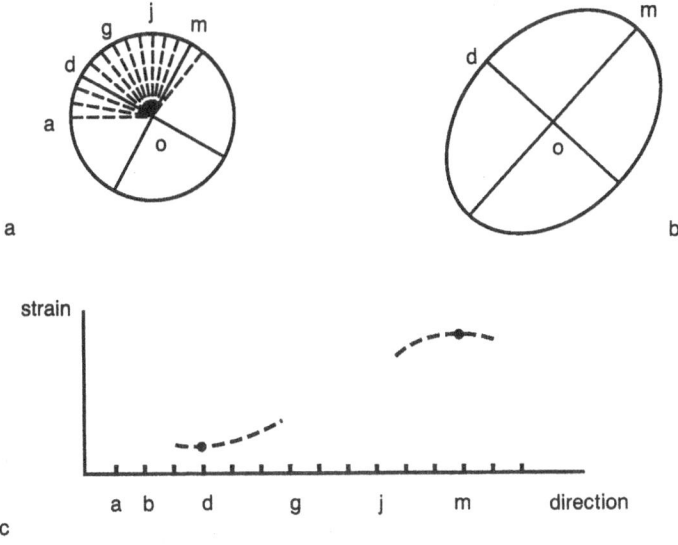

FIGURE 2.10

Question 2.17 Proof: Prove trigonometrically the assertions just given.

Summary   The manner in which the linear strain varies with direction (in a two-dimensional sample in which the strain is homogeneous) can be shown in three equivalent ways:

1. by a circle plus ellipse;
2. by a pair of numbers, (x,y);
3. by a sinusoidal profile.

In three dimensions, equivalents are

1. a sphere plus ellipsoid;
2. a trio of numbers;
3. three separate sinusoidal profiles in the three principal planes.

The marvelous terseness of the second form is worth noting. The next topic is shear strains, about which vast amounts of writing exist. Shear strains are interesting and sometimes convenient to imagine, but the idea is totally unnecessary; in some ways, mechanics would have been simpler if shear strains had never been invented—the sphere-plus-ellipsoid and its three-number specification are the kernel of the matter. On the other hand, the idea of a shear strain is sometimes a great convenience and the effect sometimes pleasing to the eye.

SHEAR STRAIN

The essential relation of linear strain to shear strain is shown schematically in Figure 2.11. If linear strain is thought of as a ratio m/l, where l is the length of a line and m is a displacement of one end parallel to the line's length, then **shear strain** is a similar ratio m/l where now the displacement m is normal to the line's length. Thus, two directions at right angles are essential for shear strain. A more formal definition is

If the angle between two lines is initially 90° and is found after deformation to differ from 90° by an angle ϕ, the shear strain is tan ϕ.

The angle itself is sometimes called the angle of shear. If the two lines are named J and K, the shear of J with respect to K is clearly equal in amount to the shear of K with respect to J, but if J shears clockwise with respect to K, then K shears counterclockwise with respect to J. These are not two different processes, they are two different descriptions of the same process; see Figure 2.12.

Geologists spend a lot of time dealing with layered materials (such as bedded sediments), and it is natural to think of a feature such as perhaps a worm tube shearing with respect to the layering (see Figure 2.13). It is a good habit always to take the second point of view as well, and to think of the bedding planes being sheared with respect to the worm tube.

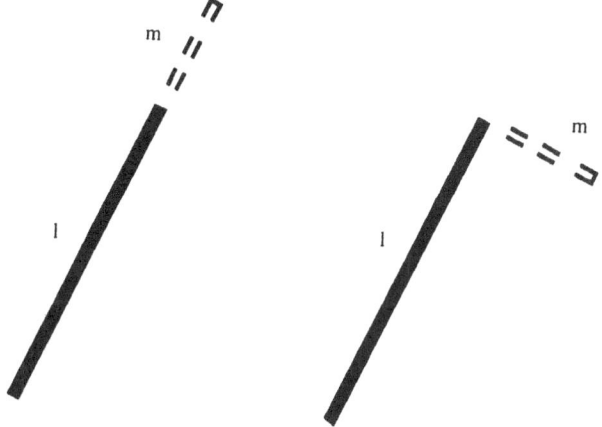

FIGURE 2.11

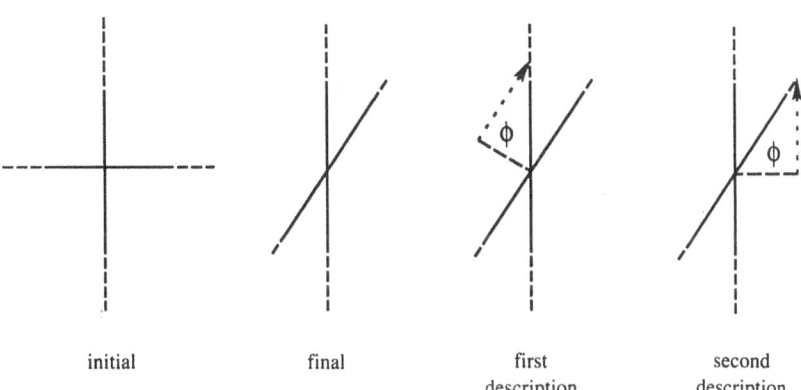

initial   final   first   second
                              description   description

FIGURE 2.12

FIGURE 2.13

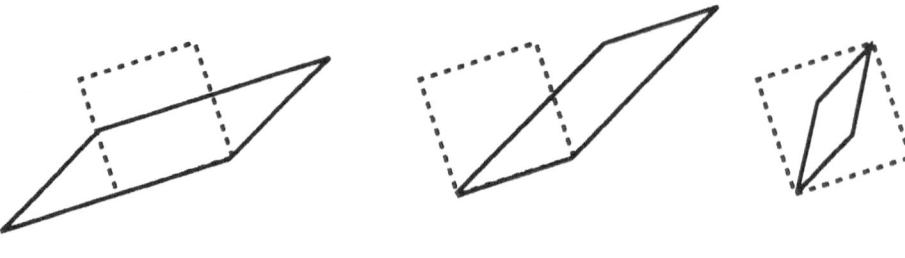

FIGURE 2.14

Question 2.18 Shear strain examples: Make diagrams to represent a shear strain of 0.2 and a shear strain of 2.

Note that in answering Question 2.18, it is natural to draw a line of length L and a line of length $2L$ at right angles to it, and to complete a triangle with a line of length $L\sqrt{5}$. But a shear strain of 2 *does not* imply that any line has elongated from L to $L\sqrt{5}$. A variety of changes of shape, each of which involves the sides of a square suffering a shear strain of 2, is shown in Figure 2.14.

Distinction from Rotation When *lines* are under discussion, one can say that line J rotates with respect to line K, even if J and K never make a right angle. If J and K initially make a right angle, to say that J rotates with respect to K through angle ϕ is equivalent to saying that J shears with respect to K by an amount tan ϕ. But suppose that, for once, we imagine an external reference frame and that, of two lines initially at right angles, one rotates 20° with respect to the frame and the other rotates 30° as in Figure 2.15. The amount of shear is definitely tan 10°, but can we give a clear, simple statement of the amount of rotation?

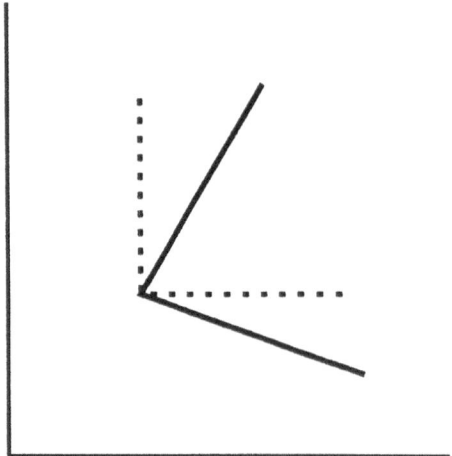

FIGURE 2.15

Remember that the entire discussion concerns lines of particles in a homogeneously deformed material, so that the deformation must have principal directions. Along these, we find lines of particles that make 90° after deformation and made 90° initially—i.e., lines of particles that have suffered no shear. The rotation of the principal directions with respect to an external frame can be stated, e.g., in degrees or radians and, if one wishes to specify amounts of rotation, this is the best quantity to focus on.

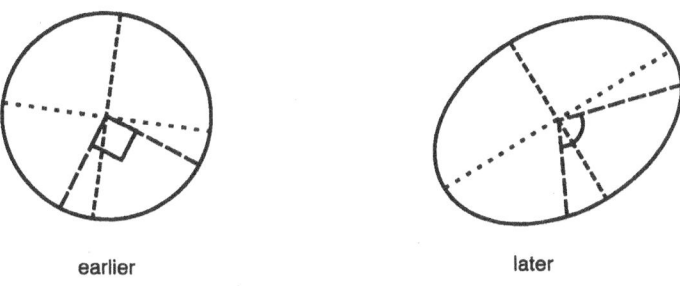

earlier   later

FIGURE 2.16

Question 2.19 Distinguishing shear strain from rotation: Refer to Figure 2.16.

(a) How much rotation has affected the sample as a   (b) How much shear strain has affected the pair of
    whole?   dashed lines?

Variation of Shear Strain with Direction

The amount of *rotation* exhibited in Figure 2.16 is a single amount that affects the entire sample, but the amount of shear strain is an attribute of just the particular pair of dashed lines shown. As already noted, the principal directions suffer no shear strain, and a series of pairs of lines can be imagined, as in Figure 2.17.

If we think of the principal directions as the first of such a series, with shear strain of 0°, then pairs i–I, ii–II, and iii–III show successively larger amounts of shear strain; but beyond example iii (where the two dashed lines are symmetrically disposed with respect to the ellipse axes) further examples such as iv are simply reflections of possibilities we have already considered; for example, with the long axis as mirror, diagram iv is more or less a reflection of diagram ii. The trend of the shear strain magnitudes can be displayed if we use again an axis like the one in Figure 2.10 (see Figure 2.18).

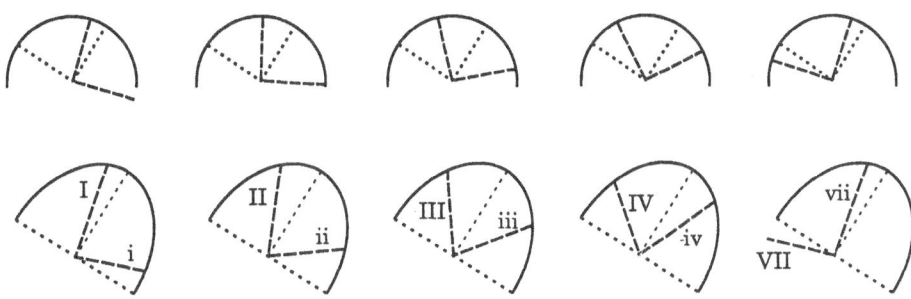

FIGURE 2.17

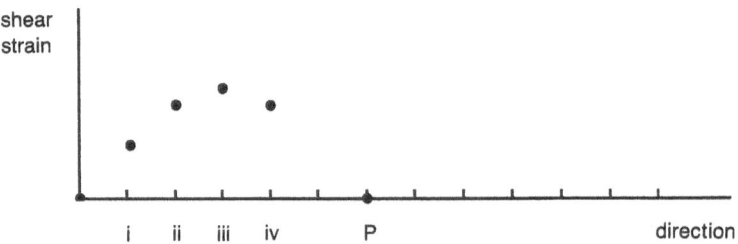

FIGURE 2.18

By the time line v or vi in the series coincides with the ellipse's long axis, the associated shear strain is down to zero again (point P in Figure 2.18). Beyond this stage, again, we only come to possibilities that repeat behaviors we have already graphed: for example, dashed lines vii and VII are a repeat of dashed lines i and I, except that now we are giving attention to an obtuse angle rather than the corresponding acute angle. The deviation from 90° is necessarily the same for the pair vii, VII as it was for the pair i, I. It is customary in continuing the profile in Figure 2.18 to cross to the underside of the horizontal axis; then the complete picture of variation of shear-stress magnitude with orientation of the line-pair considered is as in Figure 2.19.

As with linear strains, so with shear strains, when they are large, the exact shape of the profile deviates from a sine wave; but if we treat shear strain *rates,* or *small* strains, the profile becomes a sine curve. Even more conveniently, if we plot on the vertical axis *one-half* of the shear strain, the shear-strain profile becomes identical with the linear strain profile except for being displaced 45° horizontally and being symmetrical above and below the axis (whereas the linear strain curve can be wholly above the axis or wholly below it, etc.) An example of a composite diagram is given in Figure 2.20.

Maximum Shear Strain   If the two curves in Figure 2.20 were identical, a corollary would be

$$\text{maximum shear strain} = e_{max} - e_{min}, \tag{2.16}$$

where e_{max} and e_{min} are the maximum and minimum values of linear strain. As just noted, the relation is exact for strain rates or small strains, but neither of these can be conveniently explored with simple diagrams. Strains shown by diagrams have to be big enough to be seen, and in this condition the relation given is not exact but is worth exploring all the same.

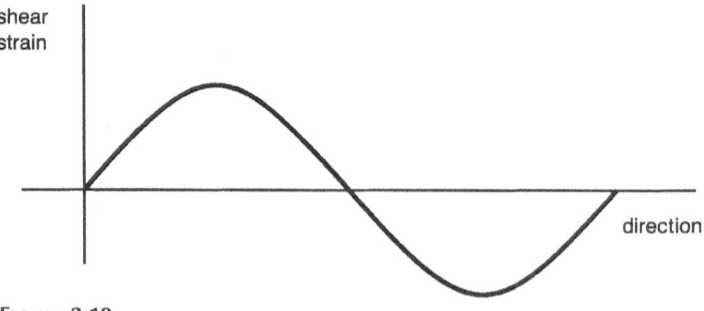

FIGURE 2.19

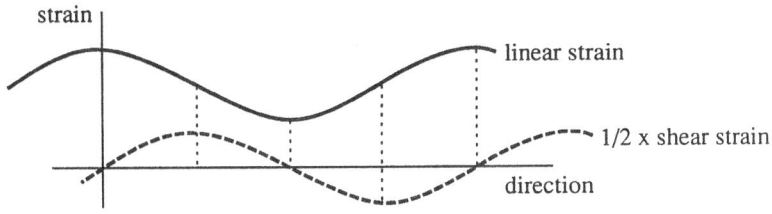

FIGURE 2.20

Question 2.20 Shear strain and linear strain: In Figure 2.21, estimate

(a) the linear strain e_x of the shape's top edge,
(b) the linear strain e_y of the shape's side, and
(c) the deviation β of angle α from 90°.

How close is tan β to difference $e_x - e_y$? (If it is not close, check the signs you have given to e_x and e_y; throughout the book, we treat elongations as positive and shortenings as negative strains.)

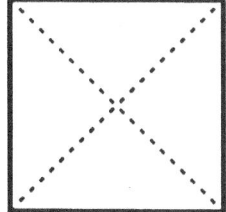

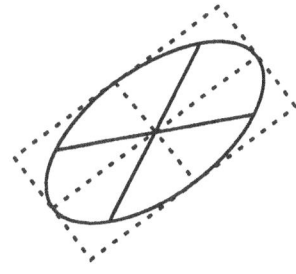

FIGURE 2.21

Question 2.21 Shear strain and linear strain, continued: Repeat Question 2.20 for two or three more rectangles of your own invention, until you are satisfied that for strains less than 0.5 the result is roughly true but is not exactly true.

A point to note about Figure 2.21 is that we have switched back from circle-and-ellipse diagrams to a rectangle diagram; the virtue is that the diagonals of a square are so easy to see and follow as they deform. The lesson to learn is that both kinds of diagram are useful: for every rectangle, imagine an ellipse inside, and for every ellipse, imagine a rectangle around it—this kind of double vision helps keep track of the many aspects of strain that go on simultaneously (see Figure 2.22).

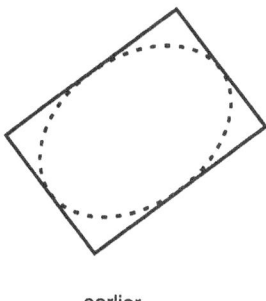

earlier

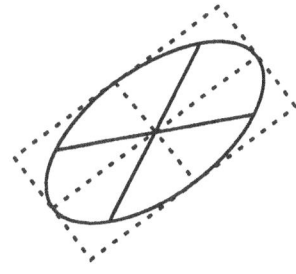

later

FIGURE 2.22

A second point to note about Figure 2.21 is the assumption involved in Answer 2.20 that the diagonals of the initial square lead to the *maximum* shear suffered by any pair of initially perpendicular lines. The sequence in Figure 2.17 shows that this must be so— the *maximum* numerical value and the *symmetrical* diagram iii have to go together.

FIGURE 2.23

Question 2.22 Maximum shear strain: All the fossils in Figure 2.23 have suffered the same strain. What is the maximum shear strain suffered by any pair of lines in the deformation? (*Note:* There is a tedious way of reaching an answer and a quick way.)

Question 2.23 Maximum shear strain, continued: The preceding text suggests that if the maximum elongation is 0.020 and the minimum elongation is 0.008 (difference 0.012), then maximum shear strain = 0.012. Verify this by considering a square with edge 1 m and the displacement of its corners when the square deforms, assuming that its center stays fixed. By how many millimeters will each corner move, horizontally and vertically? Through what angle will the square's diagonal rotate? (Consider first the diagonal going to the top right corner and then the diagonal going to the top left corner. Make a large diagram to show how the top right corner moves. Use the fact that, for a small angle α, $\tan \alpha = \alpha = \dfrac{\text{length of a short arc}}{\text{length of a long radius}}$.

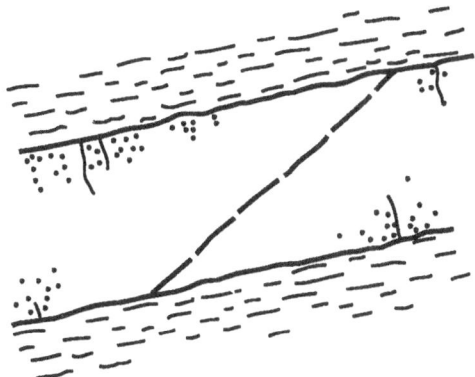

FigURE 2.24

Question 2.24 Shear strain in a general case:

(a) Consider a passive marker line at 30° to bedding in a layer of rock, Figure 2.24.

Take axes Ox and Oy parallel and perpendicular to the base of the layer; let the shortening parallel to Ox be 0.04 (strain $= -0.04$) and elongation parallel to Oy be 0.02. Through what angle does the passive marker rotate?

(b) Through what angle does a second marker perpendicular to the first one rotate?

(c) How much shear strain is suffered by the pair of marker lines?

Question 2.25 Shear strain in a general case, continued:

(a) Use the same line of reasoning to show that if the principal strains are e_x along Ox and e_y along Oy, a line of length L making an angle θ with Ox initially will rotate through an angle $1/2\,(e_x - e_y)\sin 2\theta$, if e_x and e_y are small.

(b) Show that the rotation in (a) is half the shear strain suffered by the line considered with respect to a line initially at right angles to it.

(c) Show that if the quantity $1/2 \times$ (shear strain) is plotted, the resulting curve has a vertical separation, crest to trough, that is the same as for the linear-strain curve.

Summary

1. A set of linear strains always has a set of shear strains in attendance. See Figure 2.25: every such full line is accompanied by a dashed line.

2. The two curves are identical for strain rates or small strains: the maximum shear-strain rate equals the difference in linear-strain rates

$$\dot{\gamma}_{\max} = \dot{e}_{\max} - \dot{e}_{\min}. \tag{2.17}$$

3. For strain rates or small strains, the equations of the curves are

$$e = \tfrac{1}{2}(e_{\max} + e_{\min}) + \tfrac{1}{2}(e_{\max} - e_{\min})\cdot\cos 2\theta, \tag{2.18}$$

$$\tfrac{1}{2}\gamma = \tfrac{1}{2}(e_{\max} - e_{\min})\cdot\sin 2\theta. \tag{2.19}$$

4. It continues to be true that the entire picture of all the strains in the straining neighborhood is contained in just the two numbers, $e_{\max}$ and $e_{\min}$ (or three numbers in

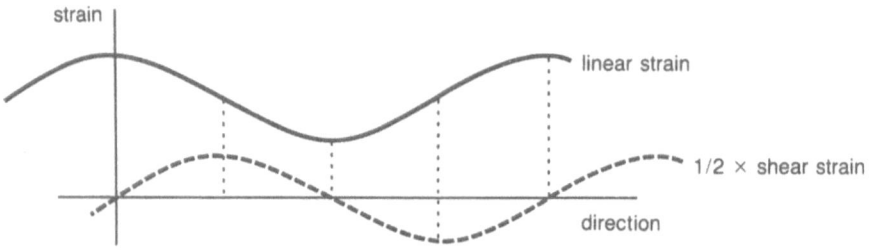

FIGURE 2.25

three dimensions). The numbers specify the strain; the geologist's imagination fleshes out the numbers with details of all the many processes that are going on simultaneously.

LAMINAR MATERIALS

Laminar materials are very common in geology and elsewhere. The characteristic feature is a layer of material that, although highly deformable, exists as a thin parallel-sided sheet, and the thickness between the sides hardly changes (see Figure 2.26). The film of grease in a simple cylindrical bearing is an important instance from technology. The deformation that occurs between parallel planar boundaries is easily imagined as a shear strain. This aspect will be explored first, and then the extent to which the process also involves linear strains and rotation will be taken up after.

Cumulative Shear Strain

Shear strains can be summed and averaged in the same manner as linear strains; Questions 2.26 and 2.27 intentionally resemble Questions 2.2 and 2.3.

	Thickness	Shear strain
Question 2.26 Average shear strain: The following stratigraphic column shows thicknesses for four formations and the amount by which each formation has been sheared. What is the average shear strain for the whole column?	80 m	0.10
	48 m	0.15
	10 m	0.66
	40 m	0.02

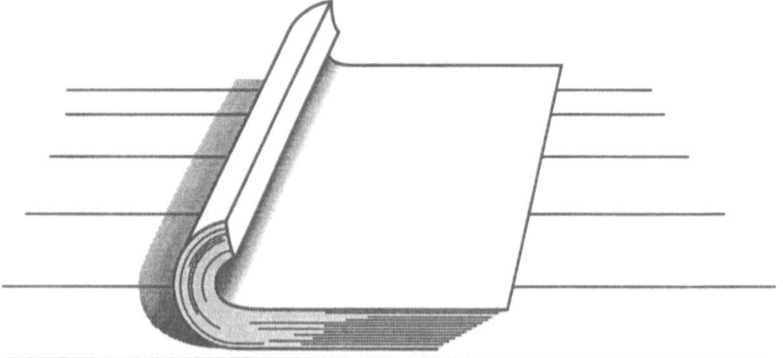

FIGURE 2.26

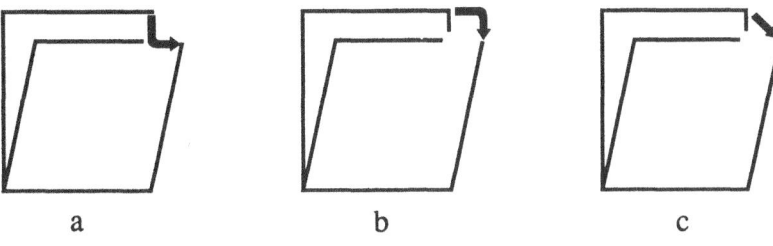

FIGURE 2.27

Question 2.27 Average shear strain, continued: A sedimentary succession is 3/5 shale and 2/5 limestone. Shear strains are 0.32 in the shale and 0.12 in the limestone. What is the average shear strain over the whole succession?
(The topic is continued in Question 5.4.)

We see that shear strains can be summed, weighted, averaged, and integrated in the same way as linear strains if the units being strained are slabs of constant thickness.

If the slabs change thickness while shearing, behavior continues to be intelligible, but we quickly lose simplicity. For example, imagine the thinning in Questions 2.2 and 2.3 to be combined with the shearing just discussed. Possible histories are shown in Figure 2.27. Three histories are envisaged: in (a) thinning almost wholly precedes shearing; in (b) shearing almost wholly precedes thinning; in (c) thinning and shearing occur steadily and simultaneously. The net result is slightly different for each of these histories and would be more strongly different if larger strains were involved. To pursue some more simple thoughts while avoiding these complications, we continue to assume that layer thicknesses stay constant.

Slip Surfaces   Some layered assemblies deform without the layers themselves shearing, by one layer slipping past its neighbor. Figure 2.28 shows this possibility. If the succession in Question 2.26 contained slip distances of 8 m, 7.2 m, 6.6 m, and 0.8 m, clearly the regional effect would be the same as if these displacements were achieved by strain within the formations.

Question 2.28 Shear strain produced by slip: If planar slip surfaces are spaced on average 6 to 1 cm, and there is 1 mm of slip on each, what overall shear strain results?

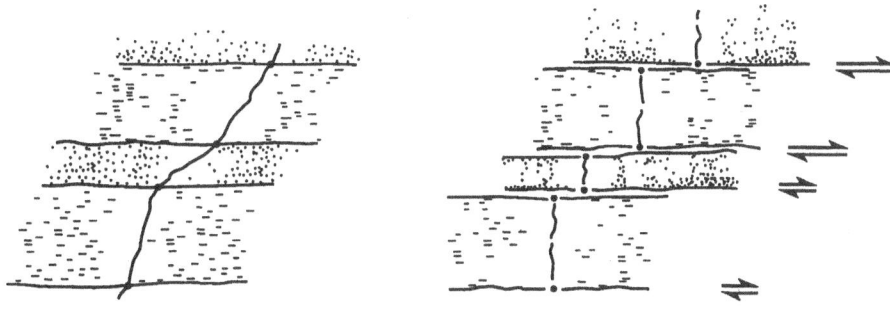

FIGURE 2.28

It is important to distinguish two concepts: In this book, **slip** designates a quantity measured in meters, millimeters, and so on, and **shear** designates the dimensionless ratio. In a volume of rock, shear and slip are alternative ways of producing displacement, or at least alternative ways for geologists to imagine that the displacement occurs. The closer one looks, the more instances one finds where what has been described as uniformly distributed shear strain turns out to be uniformly distributed slip on a set of closely spaced slip surfaces. It is perfectly permissible to use the concept of uniformly distributed shear strain all the same.

Linear Strain in Laminar Behavior

If one marks a circle on the side of a soft-covered book and then flexes the book, it is easy to see that laminar behavior can turn a circle to an ellipse. As elsewhere, in fact, if the deformation is homogeneous in the region occupied by the circle, an ellipse results, but if the deformation is inhomogeneous, the result is something more complicated.

If the book is well behaved, various amounts of deformation can be produced in succession and the following effects observed (see Figure 2.29).

At a steady shear strain rate, **1.** the material line (or line of particles) that was initially normal to the pages rotates at a rate that gradually decreases, and elongates at a rate that gradually increases. If the radius of the initial circle is R m and the displacement rate of the crest point is D m/sec, the initial rate of rotation is D/R radians per second, decreasing to zero ultimately; the rate of elongation is zero initially and D m/sec ultimately; the elongation strain rate is zero initially, climbs to a maximum and diminishes to zero ultimately as the elongation of D m/sec is distributed over an infinitely long line.

2. The long axis of the ellipse first appears at 45° to the pages and rotates, but always more slowly than the line in paragraph 1, so that the angle between them diminishes. (See bottom row in Figure 2.29.)

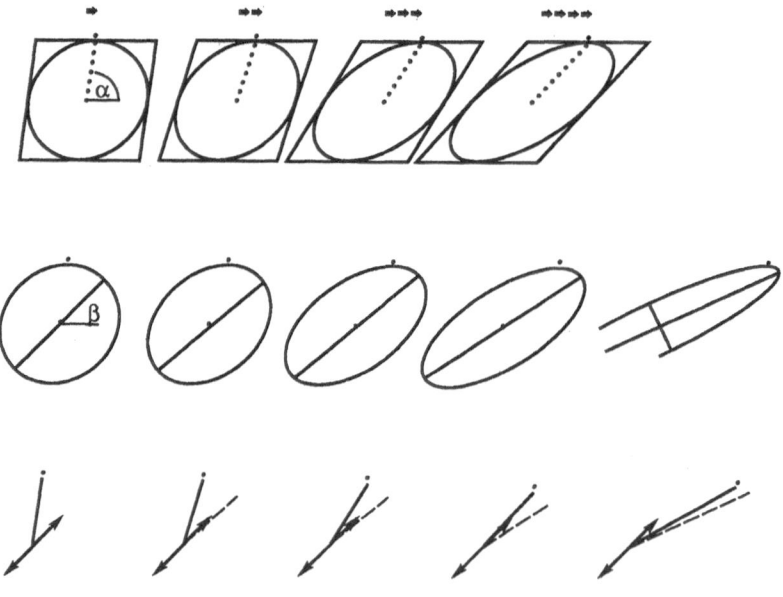

FIGURE 2.29

3. The line of particles that formed the long axis initially has its tip at $R/\sqrt{2}$ above the centerline, and its tip stays at this height throughout the deformation. Thus with respect to lines of material particles, the long axis moves away from this initially 45° line toward the initially vertical line; the long axis is said to rotate backward through the material, or at least, it fails to rotate forward as fast as the material lines rotate forward.

4. The line of particles that at any instant is elongating with maximum strain rate is always to be found at 45° to the pages; so it "rotates backward through the material" exactly as fast as the material lines rotate forward. For example, the line of particles that was initially vertical (and had zero rate of elongation) rotates forward and becomes the line of particles with maximum rate of elongation by the time its top end has moved R m and its slope is 45°.

Turning now to amounts of elongation, these are related to the amount of shear. We observe particularly the material line that was initially vertical and later has inclination α, which continues to pass through the crest point of the ellipse; and the line of greatest elongation, or current long diameter of the ellipse, at inclination β. The stretch of the first is $1/\sin\alpha$ and the stretch of the second is $1/\tan\beta$. (The first result comes directly from the diagram; the second result is proved with the help of Figure 2.30).

Although these results are geometrically rather elegant, note that the entire discussion is highly special and hinges on constancy of thickness normal to the book pages or other laminae. Any time that slipping book pages or a card deck is used as a model for deforming rocks, there is the possibility of misinterpretation creeping in, because of rocks' tendency to change thickness while slipping. Nonetheless the models certainly provide a quick and easy beginning to an exploration of how the rocks might actually have behaved.

Comparison with Coaxial Deformation   The comparison begins with Figure 2.31, which shows a marker that was initially a circle. Diagram (a) shows the effect of a small

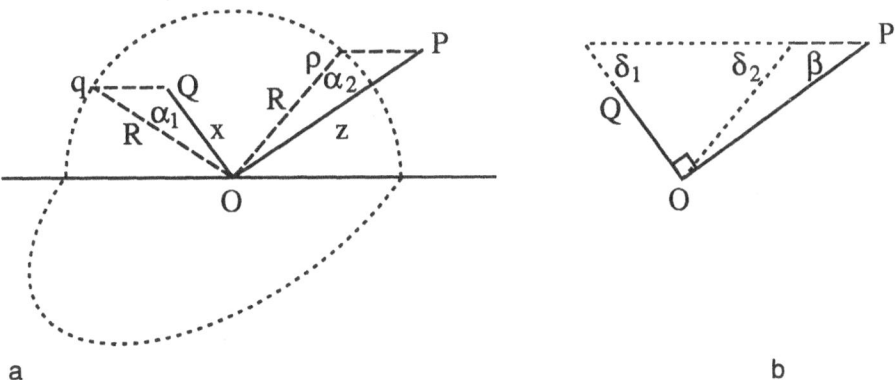

a   b

FIGURE 2.30   Let lines x and z be the axes of an ellipse produced from the given circle by shear strain. Their original positions are found by plotting horizontal lines backward from P to p and Q to q. The ellipse has the same area as the original circle so $XZ = R^2$ or $X/R = R/Z$; also $\alpha_1 = \alpha_2$ because POQ and pOq are both right angles, so the two triangles are similar. In particular OP and OQ have equal slopes and $\delta_1 = \delta_2$ in diagram (b). By the sine rule, $Z/R = \sin\delta_2/\sin\beta = \sin\delta_1/\sin\beta = \cos\beta/\sin\beta$.

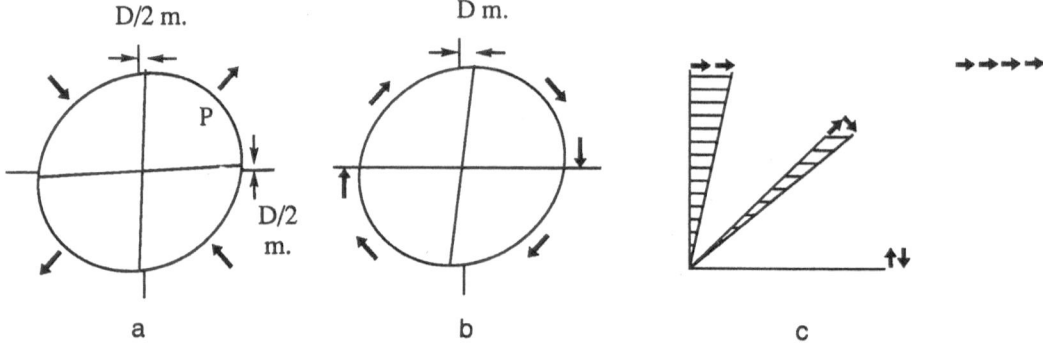

FIGURE 2.31

flattening on axes at 45° to the horizontal. Diagram (b) shows the effect of following the flattening by a small rotation. If the magnitudes of the effects are properly matched, the result can be *as if* the sample had been sheared at constant vertical thickness; as an example, the two-step path of point *P* is shown in diagram (c) and it appears that the net effect is a horizontal translation.

We now take a slightly artificial point of view and imagine an operator creating a gradually increasing deformation. She applies the small flattening, then the small rotation; then, continuing to keep the flattener oriented at 45°, she flattens again and rotates again; and then again and again. Somewhat as with extruding spaghetti, by continually working on the 45° diagonal she feeds more and more material across the top of the diagram toward the right, and the point that was initially the circle's crest-point gets driven farther and farther away. (We are assuming homogeneity, so that away to the right there is another operator with a 45° flattener who keeps the crest-point moving along.) The point of this farcical excursion is to emphasize the contrast with coaxial deformation, where the operator would omit the rotations and just keep on flattening.

To continue the comparison in quantitative terms, we note that in Figure 2.31(a), the maximum shear strain is D/R (in two rotations of $(1/2) D/R$. Hence, from Equation (2.16) $e_{max} - e_{min} = D/R$. Since we are conserving area, we need also $e_{max} + e_{min} = 0$ so that e_{max} and e_{min} both equal $(1/2) D/R$, one shortening and the other elongating. Let $(1/2) D/R$ be, as an example, 1 percent or 0.01: then 30 bursts applied coaxially would give a total strain where the stretch F is given by $\log_n F = 30.(0.01) = 0.3$ from Equation (2.4). Hence $F = 1.35$, and the elongation strain of the ellipse long axis would be 0.35.

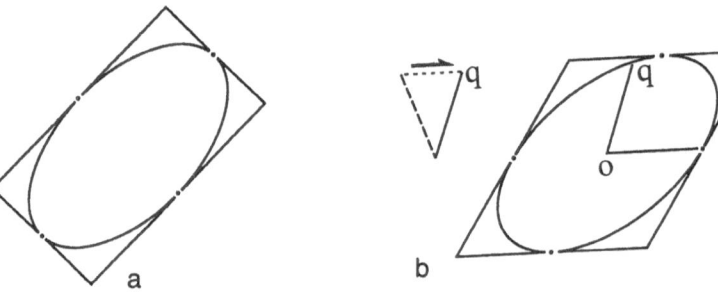

FIGURE 2.32

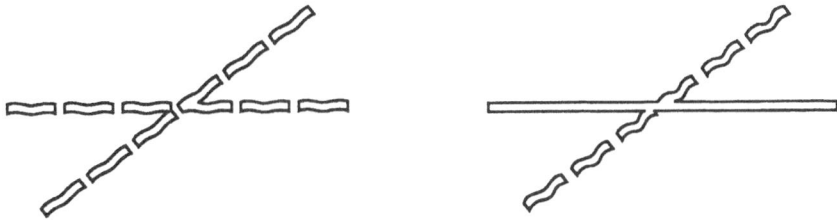

FIGURE 2.33

By contrast, 30 bursts applied with accompanying rotations would give a total strain where the crest point moves $30D$ m and the shear strain $= 30(0.02) = 0.6$. The two outcomes, Figure 2.32, are not very different except as regards the history of individual lines of particles.

The fact that in the two processes, lines of particles have different histories is best illustrated by Op and Oq. In Figure 2.32(b) the material line Op does nothing at all, while material line Oq has rotated with early shortening and later elongation; in (a), two corresponding lines both have the same history. If the sample were in fact a veined rock, with veins suitably oriented to reveal the histories, the results could be visibly different, as in Figure 2.33.

In the idealizations discussed, the differences seem small, even trivial perhaps. But it will be shown in later chapters that several processes in rock deformation are self-accentuating: in a rock that is originally only slightly inhomogeneous, processes run slightly more briskly in certain spots—but if the effect is to increase the inhomogeneity, there is positive feedback and the behavior is self-accelerating or "runaway." This kind of self-accentuating inhomogeneity has been totally excluded from the preceding discussion, but if inhomogeneities were present, they might easily pick up small differences of texture such as those shown in Figure 2.33 and develop them.

Summary

1. In coaxial deformation, there is one line of particles that suffers maximum shortening in the early stages and continues to be the line of maximum shortening throughout the deformation; the principal axes of the deformation lie along the same lines of material particles throughout the deformation.

In laminar behavior, there is one line of particles that suffers no deformation at all, and all lines parallel to this line simply translate along their own length. The principal axes of the momentary deformation rotate with respect to the material, and so one line of particles is the line of maximum shortening rate at one stage, and a different line of particles takes this role later on.

The geometrical differences, which are small in ideally smooth continuous, featureless pastes, may be sufficient to trigger significantly different changes of texture in materials such as rocks, whose behavior is sensitive to small rearrangements of the constituent grains.

2. If the material retains constant mechanical properties, the energy consumed in n bursts of small strain $(e, -e)$ is the same regardless of whether or not rotations occur during the process—that is, regardless of whether the deformation is coaxial, laminar, or other variant.

3. Laminar behavior is seen in materials that are themselves laminae (book pages, a deck of cards, etc.), but the same behavior is shown by simple materials such as water if it is confined between broad, parallel, rigid boundary surfaces.

The question arises: Can a featureless continuous material be *converted into* a set of laminae by a deformation process? This question is of great importance in geology, lubrication, and elsewhere; it cannot be attacked in a formal way until after Chapter 5 but is one suitable for brooding on, and for exercise of the imagination.

Question 2.29 Estimating strain from veins:

(a) The present shape of the deformed vein in Figure 2.33(b) suggests that it has been shortened (to give folds) and then elongated again (with separation into fragments). Measure the ratio (sum of gap widths)/(total length) for some representative portion of the deformed vein. Form an estimate of the ratio (present length)/(length when at maximum shortness).

(b) The undeformed vein perhaps shows a direction along which microlaminae have slipped in laminar flow, carrying the deformed vein with them. Is the ratio you calculated in part (a) larger or smaller than the ratio you would expect using the laminar-flow hypothesis? If there is a discrepancy, consider a possible cause or causes.

CHANGE OF STRAIN FROM POINT TO POINT

Earlier parts of this chapter relied heavily on the idea of a homogeneous strain; Figure 2.8(c) is the only illustration yet given of behavior where the amount of strain varies from point to point. But obviously in mountain belts and in many single outcrops, strain varies strikingly from one point to another. The main purpose of this chapter is to review those ideas about strain as are needed to understand the mechanical behavior of rocks, and variation of strain from point to point is not directly on the main line of advance. However, a brief discussion serves to reinforce ideas already presented, and will serve as a source of examples for later discussion of forces, stresses, and energy conversions.

Folding

Prime examples of strain varying from point to point are provided by folded layers; some idealized forms are shown in Figure 2.34. In the figure, all the layers shown have the same profile along their centerline, but have been filled out in different ways. Diagram (a) shows laminar behavior supposing a set of laminae are all parallel to line A. Diagram (b) is similar except that the laminae are imagined at a slightly different angle to the centerline profile (line B); the small change has a noticeable effect on the relative thicknesses of the fold limbs. Diagram (c) is more in the manner of flexed book pages: the top surface wraps around the centerline, maintaining a fixed perpendicular distance from it at all points; but marks are shown at uniform spacing along the top surface, the centerline and the bottom surface as they would be if these surfaces suffered no linear strain along the fold surfaces (In (a), the direction of no linear strain is parallel to line A.)

By contrast, the property in diagram (d) is no shear strain parallel to the fold surfaces. A set of lines that was normal to the surfaces before folding is shown, still normal to the fold surfaces at all points. (Regarding separation distance, (d) has the same property as (c), namely that the perpendicular separation of the fold surfaces is the same all round the fold; folds (c) and (d) have the same external shape.)

It is to be noted that folds (a) and (b) could be nested so as to fill space, but folds (c) and (d) could not. A real layered rock resists elongation parallel to its layers (behavior c)

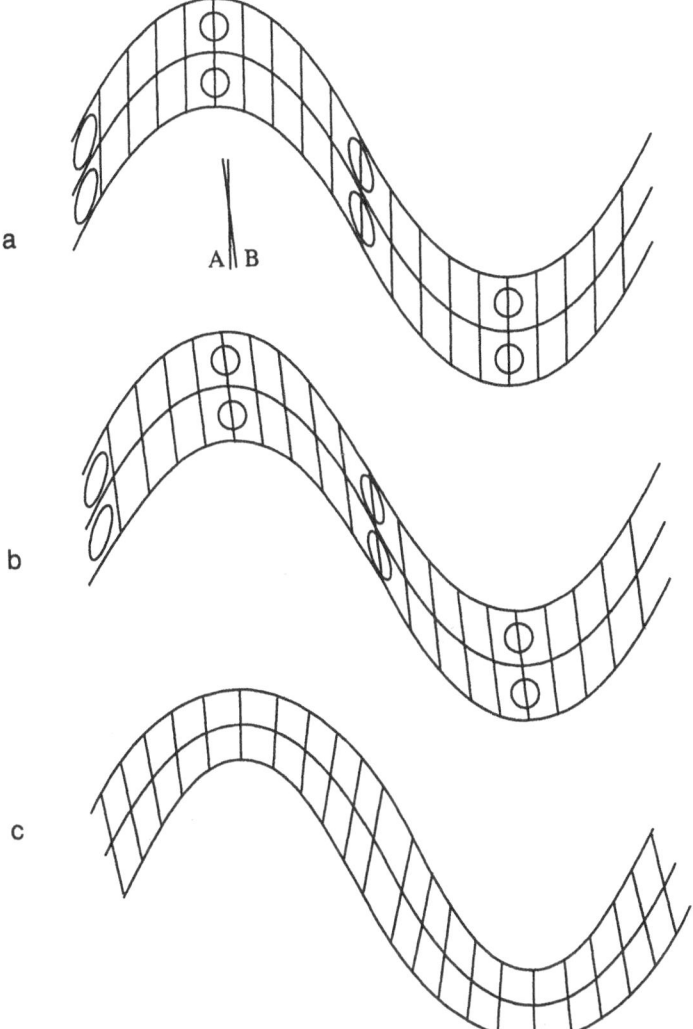

FIGURE 2.34   (Continued)

and resists shearing parallel to its layers (behavior d); it also tends to fill space without the creation of voids (behavior a or b), while having some layers a little stiffer mechanically than others. Diagram (e) suggests how the different tendencies could combine. This is a global compromise, but even (e) contains simplifying assumptions—first, that there is no slip at any bedding plane, and second, that there is no systematic change in the fold profile as one goes higher in the succession.

Points to emphasize are that all over the folds, circles are turning into ellipses and shear strains are accompanying linear strains. The recipe admits endless variations, but the ingredients are always the same: at any point at any instant, the local process is a circle turning elliptical, in a manner specified by the two principal linear strain rates, $\dot{e}_{max}$ and $\dot{e}_{min}$. The mechanics of folding, like so much of geological mechanics, has to zero in on these two.

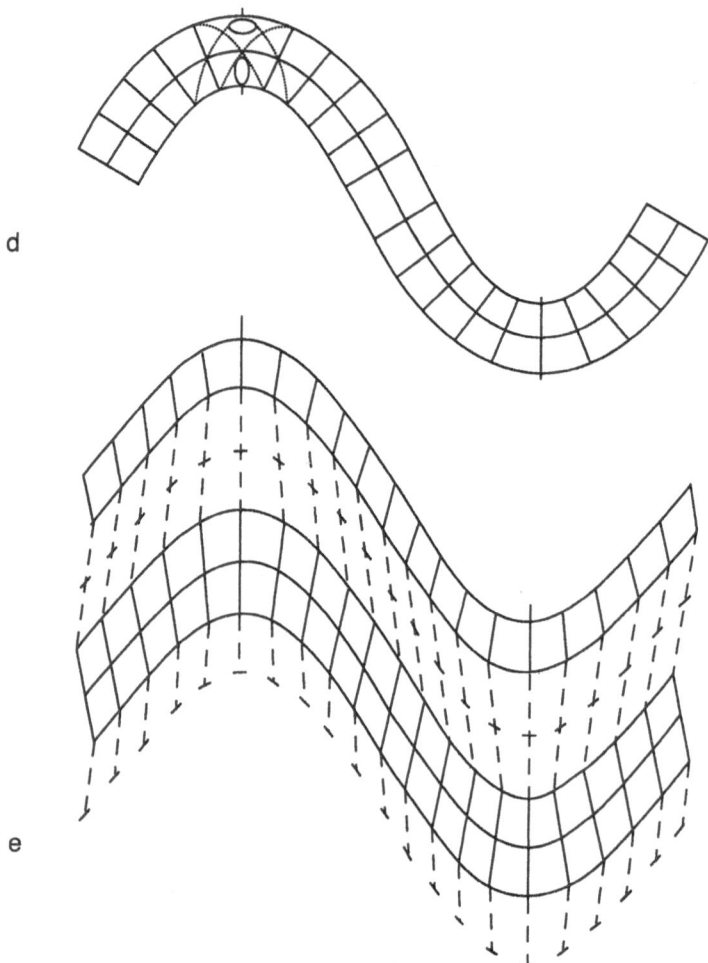

d

e

FIGURE 2.34 (*Continued*)

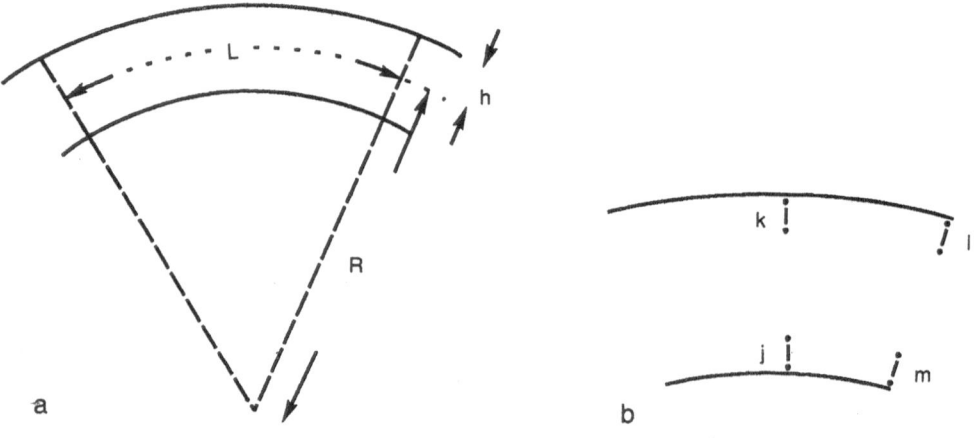

a

b

FIGURE 2.35

Circular Arcs   One of the conspicuous difficulties with folds is that the processes vary so significantly from point to point. Circular arcs are exceptional in that strain is somewhat uniform, and therefore can be concisely described.

Figure 2.35 shows a portion of a circular arc with centerline of length L and radius R. Temporarily, it is assumed that the inner and outer surfaces are uniformly at distance h from the centerline and so have radii $R + h$ and $R - h$.

Question 2.30 Strain at a circular arc:

(a) If $L = 40$ m, $R = 70$ m, and $h = 6$ m, what is the linear strain along the outer surface?

(b) What is the linear strain along the inner surface?

(c) How does the present length of the line segment j in diagram (b) probably compare with its original length (i) qualitatively and (ii) quantitatively, assuming that areas are locally conserved (i.e., squares become rectangles whose area equals that of the original square)?

(d) Repeat part (c) for line segment k.

(e) What is the maximum shear strain suffered by a pair of lines initially at right angles in the neighborhood of k?
Repeat, for the neighborhood of j.
Repeat, for the neighborhood of the centerline.
Draw X's near j and k to show the lines that suffer maximum shear strain, and the location of the acute angles and obtuse angles.
Repeat for locations l and m.

(f) If the thickness of the layer before being deformed was $2h$, so that the centerline is approximately distance h from the outer surface, what can you say more exactly about the present distance from the centerline to the outer surface? Repeat for the inner surface.

Not infrequently, the initial information does not include the radius of curvature, but is as in Figure 2.36(a) instead. Generally, for a circle, $xy = z^2$, and if g is small in comparison with the diameter $2R$ and the length of arc L, this translates to

$$R = L^2/8g. \qquad (2.20)$$

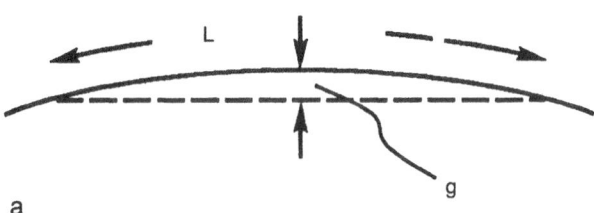

a

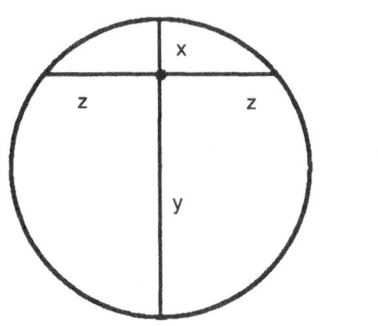

b

FIGURE 2.36

Question 2.31 Strain in a cylindrical trough: A limestone unit 800 ft thick and 50 km across is laid down horizontally and lithified; then its center is depressed 1000 ft with respect to its east and west margins to give a cylindrical trough.

(a) If the centerline of the layer does not change length, by how many feet does the top surface shorten? What is the linear strain of a east-west line on the top surface?

(b) If stylolites occurred every 10 m, how much limestone would need to be dissolved on each to accommodate the strain?

(c) If instead, a line 600 ft from the top and 200 ft from the bottom does not change length, repeat (a).

(The topic is continued in Question 2.32.)

The dimensions discussed in Question and Answer 2.31 are not outside the bounds of geological reality. It is unrealistic to suppose that, upon depression, the geometry taken up is cylindrical, with uniform curvature from point to point. More probably the curvature will vary from point to point, in regular and irregular ways. However, as a start toward the right orders of magnitude, the cylindrical assumption plays a useful role.

Summary   Parallel to the surface of a layer,

$$\text{linear strain} = h/R$$

if the surface forms a circular arc at distance h from an arc that has not changed length, of radius R, and lines initially normal to the surface remain normal to it. For a cylindrical layer, if areas are conserved, maximum shear strain at the surface = $2h/R$, suffered by planes dipping at 45° to the cylindrical surface. In Figure 2.36,

$$R = L^2/8g, \quad \text{so linear strain} \quad = 8gh/L^2.$$

Question 2.32 Strain and the earth's curvature: Reconsider Question 2.31 noting the fact that the earth's surface is curved. If two tiny islands are 50 km apart by boat, a straight line from one to the other passes through the subsurface and is shorter than 50 km.

(a) What is the maximum depth below sea level that such a line would pass?

(b) Does Answer 2.31 need to be recalculated to allow for the curvature of the earth?

Surface Waves

One more example of the way strain can vary from point to point is provided by water waves or any other deformable material with surface undulations (see Figure 2.37). If

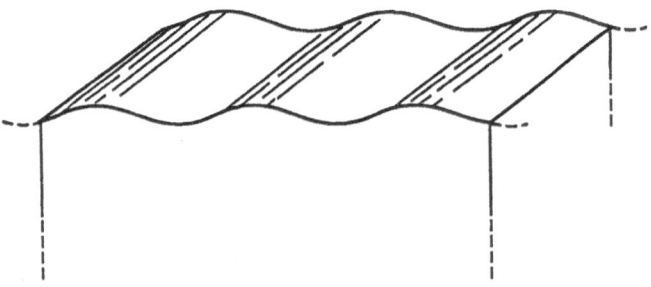

FIGURE 2.37

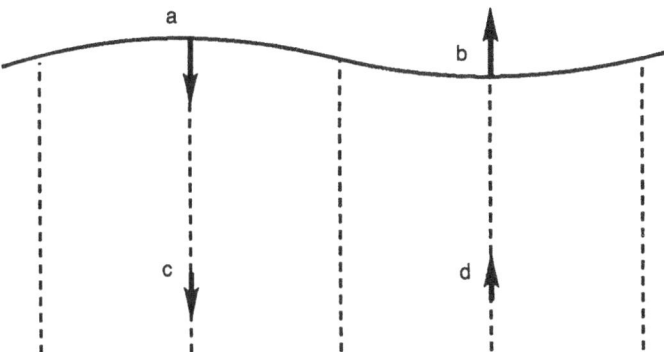

FIGURE 2.38

acted upon by only the force of gravity, how will a mobile material with such a surface respond? No geological structure is geometrically as ideal, but mountain belts and mid-ocean ridges and troughs are of this general type, so that useful geological ideas can be developed by looking at this idealized pattern.

Probably the easiest thing to imagine is a particle's displacement or velocity. Surely, under gravity, the crests will get lower and the troughs will rise, as in Figure 2.38. But these downward and upward velocities do not affect the material deep below the surface; even at points such as (c) and (d), the downward or upward velocities must be smaller. This means that distance ac is getting shorter and bd is getting longer; initial circles beneath point a are in process of flattening and becoming wider, while initial circles beneath point b are becoming narrower and taller (Figure 2.39). This means in turn that particles in locations such as e and f have horizontal velocities; these horizontal velocities must also be smaller at deeper points, so that if a line eg is vertical at one moment, it must lean to the left at later moments.

The entire process is indicated in Figure 2.40. We see that beneath point a, there is vertical shortening and downward motion; beneath point b, there is vertical elongation and upward motion; around point f, there is horizontal and vertical **shearing,** with horizontal motion and shortening of lines at about 45° to horizontal. In short, there are many separate aspects to the material strains, but all the separate aspects link together compatibly in a behavior one can readily imagine.

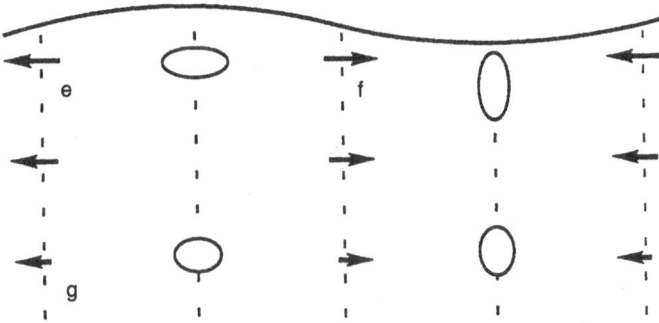

FIGURE 2.39

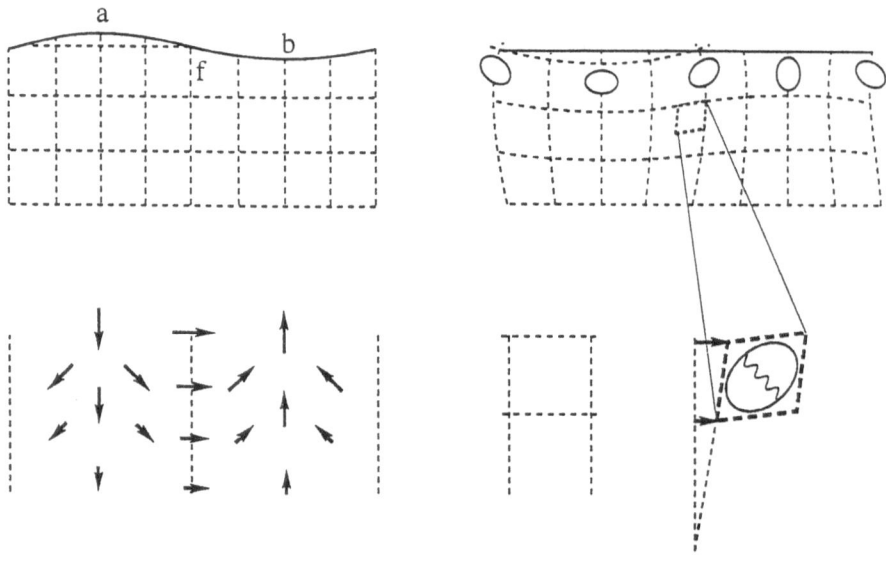

FIGURE 2.40

THE MOHR CIRCLE AND PARTITION OF STRAIN

The purpose of this section is not to introduce any more details about strain, but to introduce two fresh ways of conceiving the ensemble. We shall focus on strain rates and small strains, thus turning our backs on anything that can be readily seen, and it is worth a moment to see how these parts fit together.

Except for surficial processes, like sediment movement on a river bed, we do not see processes in geology; we mostly see results. Particularly with strains and displacements, it is the final configuration that is available for study. From this, there are two questions to ask: Can we discover the processes or history? And, can we *understand* the processes, finding causes for the effects? The first question has created the field of strain analysis, and its sister strain synthesis, not yet so far advanced. These fields are mainly geometrical: What markers reveal strain? What errors are involved? Do two markers reveal different parts of the total strain? Are the indications from one marker compatible with indications from another? And, coming to strain synthesis, if we have analyzed the strain in several localities, what principles should guide us in constructing a unified description for the region in which the localities are embedded? As noted, much of this work is geometrical and can be pursued without consideration of force magnitudes.

On the other hand, if we want *causes,* forces and stresses are one essential part of the answer, and material properties are the other part. The combination of a stress and a material of certain properties gives a response—the cause produces an effect. But the effect is either a small strain or a strain rate (or both). The material property relates behavior at some moment to the stress at that moment; there is no theory of mechanics that relates the final configuration the outcrop displays to any cause (force or stress). Thus, strain analysis and synthesis relate small strains and strain rates to accumulated observable end-results. By contrast, mechanics relates small strains and strain rates to their causes, and the two do not overlap much. An example is Figure 2.34, which skirts the topic of strain distribution in folds, a topic very rich in interest and appeal. But that belongs with strain analysis and synthesis; to progress with mechanics, we have to go back to something much simpler, like a single circle changing slightly toward becoming an ellipse.

The Mohr Circle

Essential results are repeated:

$$\dot{e} = \tfrac{1}{2}(\dot{e}_{max} + \dot{e}_{min}) + \tfrac{1}{2}(\dot{e}_{max} - \dot{e}_{min})\cdot\cos 2\theta \qquad (2.18)$$

$$\tfrac{1}{2}\dot{\gamma} = \qquad\qquad \tfrac{1}{2}(\dot{e}_{max} - \dot{e}_{min})\cdot\sin 2\theta \qquad (2.19)$$

Mohr's contribution was to put forward a diagram as in Figure 2.41, using the fact that if $\dot{e}$ and $(1/2)\dot{\gamma}$ are given by these equations, they are coordinates of some point on the circle whose center is at the point $(1/2)(\dot{e}_{max} + \dot{e}_{min})$, 0 and whose radius is $(1/2)(\dot{e}_{max} - \dot{e}_{min})$. Figure 2.41 in fact reproduces all the ideas from Figure 2.25 in even more compact form.

Question 2.33 Representation of strain by a Mohr circle:

(a) Draw a Mohr circle for the strain process defined by $\dot{e}_{max} = 8/\text{yr}$, $\dot{e}_{min} = 2/\text{yr}$. Is the process an overall expansion, overall shrinkage, or combination of some elongation and some shortening? Can you *visualize* the process, and if so, in what kind of material?

(b) According to the equations and Figure 2.25, a line at 45° to the line of greatest elongation rate is the line of greatest shear strain rate. What point on the circle represents this line?

(c) For the line with greatest shear strain rate magnitude, what is its elongation rate in Figure 2.25? Does this agree with the point you located on the circle in part (b)?

The complete equivalence of the Mohr circle, the sine-wave diagram, and the circle-plus-ellipse representation is shown in Figure 2.42. Here the large quarter-circle has radius equal to one unit of length; then the length j is a measure of OP's elongation and k is a measure of half its shear displacement (with respect to a line initially at right angles). The same lengths j and k reappear in the Mohr diagram, Figure 2.42(b), and in the sine-wave diagram 2.42(c). The angle that OP makes with the line of maximum elongation appears directly in diagram (c); in diagram (b), as the equations show, it is the *double angle* 2α that appears—the radius from the Mohr circle's center to the point with coordinates (j,k) makes 2α with the horizontal axis.

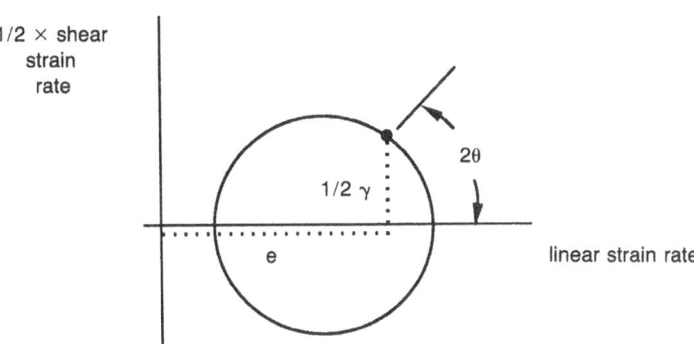

FIGURE 2.41

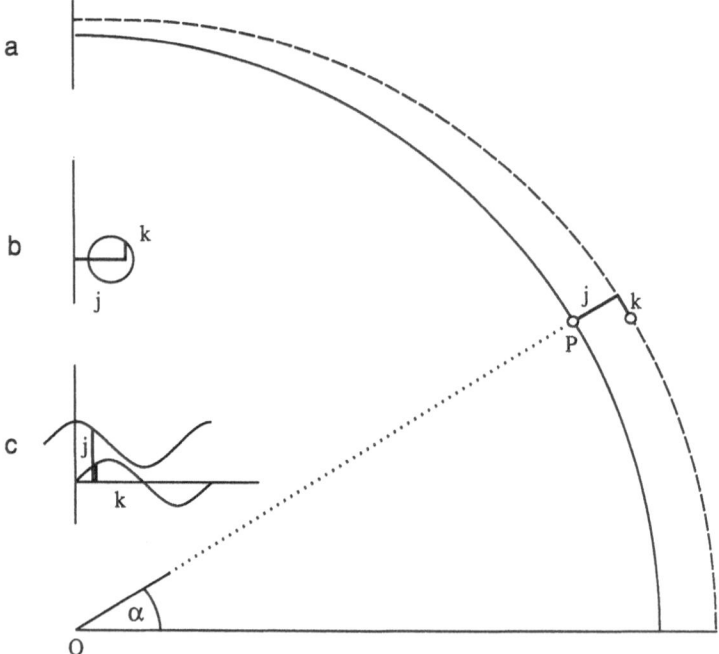

Figure 2.42

Question 2.34 Mohr circle: a second example: Draw a Mohr circle (with radius at least 6 cm) for a strain where $e_{max} = 0.02$ and $e_{min} = -0.04$. Use this circle to repeat Question 2.24:

(a) What radius in the Mohr circle is associated with e_{min}? Mark a point on the circle at the tip of this radius and label it "bedding."

(b) Mark a point on the circle 180° from the point marked "bedding" and label it "normal to bedding."

(c) Mark a point on the circle 60° from the point marked "bedding" and label it "30° to bedding" (persisting with the rule that an angle α in the real world corresponds with an arc 2α in the Mohr diagram).

(d) Find an angle of 60° in Figure 2.24 and a corresponding arc of 120° in the Mohr circle.

(e) For the point labeled "30° to bedding," read off $1/2$(*shear strain*) and *linear strain* and compare with Answer 2.24.

Working with a Mohr circle is like playing a piano: with practice, one can become amazingly nimble and deft. But for present purposes, it is the *nature* of the diagram that we wish to focus on. Because it is a circle, a natural way to describe it is in terms of its center and radius, and this leads us to a very useful fresh view of what happens when a portion of material strains.

Change of Area, Change of Shape

Equations (2.18) and (2.19) can be rewritten:

$$\dot{e} = Q + R \cos 2\theta; \tag{2.21}$$

$$\tfrac{1}{2}\dot{\gamma} = R \sin 2\theta. \tag{2.22}$$

The quantities R and Q are, respectively, the radius of the circle and the coordinate of its center on the horizontal axis. But also, R measures the *difference* $(1/2)(\dot{e}_{max} - \dot{e}_{min})$, while Q measures the *average* $(1/2)(\dot{e}_{max} + \dot{e}_{min})$. It was emphasized that the whole strain at a point (at least in two dimensions) can be summed up in just two numbers, e_{max} and e_{min}, as in the summary on page 17. We now consider that the two numbers Q and R may be just as effective, or perhaps a more effective pair.

To see the merit of the pair Q, R, consider the *processes* of change of area and change of shape—or for real geological materials in three dimensions, change of volume and change of shape. They can have different causes: on heating, a rock changes volume but need not change shape; or a waterlogged mud can change shape at constant water content, but the main way to change its volume is to let water in or out. Another way to reinforce the difference is to focus on a material's density: change of shape does not involve change of density, but change of volume does. From all these points of view, we conclude that Q is controlled by one set of processes and R is controlled by another. Remembering that we hope eventually to find *causes* for the strains described in this chapter, to describe the strains in terms of e_{max} and e_{min} is just as effective in a purely numerical sense, but it is fruitless to expect to find one cause for e_{max} taking its special value, and a separate cause controlling the value of e_{min}. As regards pinning causes to effects, we obviously have a better prospect of success if we focus on Q and R.

To consolidate these ideas with a specific example, return to Figure 2.42, where e_{max} = 0.1 and e_{min} = 0.025. These are easier to handle in units of 1/80, where e_{max} = 8/80 and e_{min} = 2/80; then the mean 5/80 = Q, and the half-difference = 3/80 = R. Also, in the figure it can be seen that the overall change is equivalent to two operations—a uniform enlargement by the factor Q, to a new circle 5/80 bigger than the original, plus a shape-change by the factor R at constant area, elongating the circle by 3/80 on one principal diameter and shrinking it 3/80 on the other (see Figure 2.43). As remarked in the preceding paragraph, in the real world we would look for one cause for the change Q, and quite probably a different cause for the effect R.

Change of Volume   In the preceding discussion, change of volume was used to emphasize that a certain set of processes is involved (that are not involved in a constant-volume change of shape), but we continued to refer to a *pair* of numbers, which is the specifier for a two-dimensional change. As stated earlier, a circle going to an ellipse in two dimensions is specified by two numbers, whereas a sphere going to an ellipsoid in three dimensions involves three numbers (or one can just as well think of a cube converting to a brick, as in Figure 2.44). Generating Q and R from e_{max} and e_{min} is straightforward, but what is the corresponding operation when we have e_{max}, e_{min}, and e_{int}?

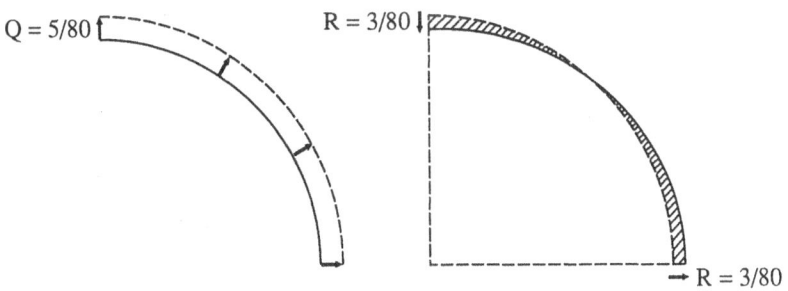

$Q = 5/80$ $R = 3/80$ $R = 3/80$

FIGURE 2.43

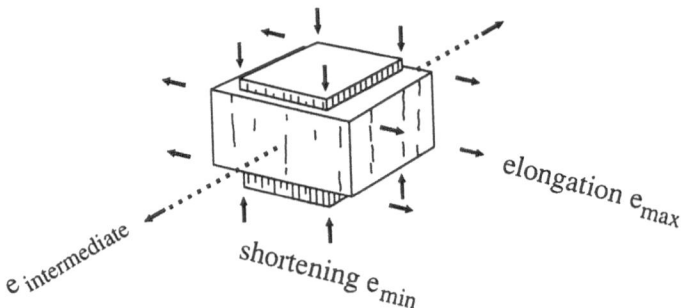

FIGURE 2.44

Consider the numerical example:

$$e_{max} = 18/1000, \qquad e_{int} = 14/1000, \qquad e_{min} = 7/1000$$

(all elongations), compared with the two-dimensional example

$$e_{max} = 8/80, \qquad e_{min} = 2/80.$$

In two dimensions, mean value = (8 + 2)/2 in eightieths, and
remainders = 8 − 5 = 3
and   2 − 5 = −3.
In three dimensions, mean value = (18 + 14 + 7)/3 in thousandths, and
remainders = 18 − 13 = 5,
14 − 13 = 1,
and   7 − 13 = −6.

(In two dimensions, the *two* remainders are always equal in magnitude so that one symbol R is sufficient, but in three dimensions the three remainders have no convenient abbreviation.) In the two-dimensional example, 5/80 is the radial change from circle to larger circle, and 3/80 is the extra bit—the change of *shape* at constant area. Correspondingly, in three dimensions, 13/1000 is the radial change from sphere to larger sphere, and +5, +1, −6 is the extra bit—the change of *shape* at constant volume.

Question 2.35 Separating out the mean strain: Separate the following into a spherical change of volume plus a constant-volume change of shape. For the first effect, give the radial strain, and for the second effect, give three axial strains that total to zero.

(a) 15/1000,   14/1000,   4/1000
(b) 15/1000,   4/1000,   −7/1000
(c) 2/1000,   −6/1000,   13/1000
(d) 3/1000,   −8/1000,   −16/1000

In Answer 2.35, part (b) turned out well: after separating out the spherical part (4/1000 in radius, or 12/1000 in volume), the remainder was a change of shape we can imagine very easily, a plane strain. The next thing to realize is that any of the other sets of remainders can be expressed as two plane strains, thus

(a)   +7,   0, −7   or   +4,   0, −4   or   . . .
       −3, +3,   0   0, +3,   −3

(c) $\begin{array}{ccc} 0 & -9 & +9 \\ -1 & 0 & +1 \end{array}$ or $\begin{array}{ccc} 0 & -10 & +10 \\ -1 & +1 & 0 \end{array}$ or ...

and so on. The *smallest* plane strains that add up to the total strain are of interest. To discover these, one ignores whichever of the remainders is farthest from zero; if the two remaining remainders are R_1 and R_2, plane strains R_1, $-R_1$, 0 and R_2, $-R_2$, 0 can always be arranged to add up to the desired overall change of shape. In the preceding example, the sets underlined are as discussed.

The upshot is that the separation of a two-dimensional strain into a uniform radial effect Q and a plane-strain change-of-shape effect R contains all the ideas we need. In three dimensions, we continue to use a radial strain Q to take care of change of volume, and represent the constant-volume change-of-shape effect by *two* plane strains R_1 and R_2. R_2 is fixed by the difference $e_{int} - e_{mean}$ and becomes zero when e_{int} is halfway between e_{max} and e_{min} as in Question 2.35(b) (see also the Pyrenees question, Question 2.11, and Figure 2.4).

Summary A strain in two dimensions can be represented by a Mohr circle. The strain can be partitioned into a radial effect (change of area with no change of shape) plus a change of shape at constant area. The change of shape correlates with the Mohr circle's radius R, and the change of area correlates with the coordinate Q of the Mohr circle's center.

In three dimensions, a strain can be partitioned into a radial effect (change of volume with no change of shape) plus a change of shape at constant volume, and the latter can be partitioned into two plane strains. The separation of change of volume from change of shape at constant volume is useful because they may have separate physical mechanisms. (Change of volume is discussed again at the end of Chapter 5.)

ANSWERS

Answer 2.1

	Initial length	Final length	Change	Strain
ab	89 units	123 u	+34 u	0.38
bc	38	43	+5	0.13
ac	127	166	+39	0.31
dc	120	103	−17	−0.14

Note: The numbers entered will depend on the measurement unit used—mm or 1/100-in. or similar. But the same strain estimates should appear whatever the measurement unit used.

Answer 2.2

Thicknesses removed by solution are

8.0 + 7.2 + 6.6 + 0.8 = 22.6 m total removed.

Therefore,

Average thinning ratio = 22.6/(initial total thickness)
= 22.6/178 = 0.127...

Answer 2.3

One needs arbitrarily to decide how much shale and limestone to consider—for example, a total thickness of T m.

Then original thickness of shale $= 2T/5$ m, and
$$\text{loss of shale} = 0.12 \times 2T/5 \text{ m};$$

original thickness of limestone $= 3T/5$ m, and
$$\text{loss of limestone} = 0.18 \times 3T/5 \text{ m};$$
$$\text{total loss} = 0.156T \text{ m (from } T$$
$$\text{m initial).}$$

Therefore, loss-fraction $= 0.156$.

Answer 2.4

(a) Overall return $\bar{r} = \dfrac{\sum\limits^{n} r_i s_i}{\sum\limits_n s_i}$.

(b) Mass $= \displaystyle\int^{V} \gamma(v) \cdot dv;$ Average density

$$= \frac{\displaystyle\int^{V} \gamma(V) \cdot dv}{\displaystyle\int^{V} dv} = \frac{\displaystyle\int^{V} \gamma(v) \cdot dv}{V}.$$

(c) Density $= \displaystyle\sum^{n} \gamma_i f_i.$

Answer 2.5

At start of eighth minute,
$$\text{gradient} = 0.33 \text{ cm/min},$$
$$\text{length} = 8.1 \text{ cm},$$
$$\text{strain rate} = \frac{0.33}{8.1} = 0.041 \text{ per min.}$$

At end of eighth minute,
$$\text{gradient} = 1.0 \text{ cm/min},$$
$$\text{length} = 8.7 \text{ cm},$$
$$\text{strain rate} = \frac{1.0}{8.7} = 0.115 \text{ per min.}$$

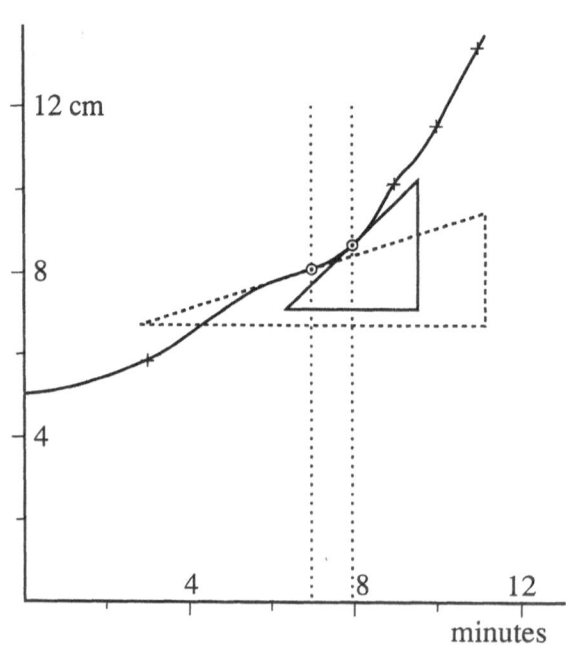

FIGURE 2.45

The arithmetic average of these, 0.078 per min, is close to the average calculated already, 0.074 per min, but one need not expect the two to agree exactly.

Answer 2.6

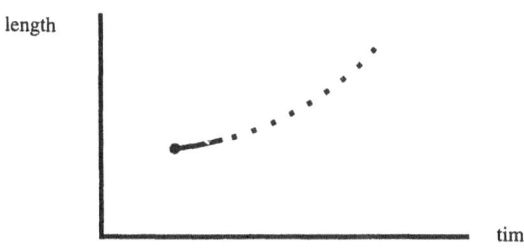

length

time

FIGURE **2.46**

(a) A plot of length versus time would have a positive slope, but would also have an upward curvature, because the length (on which the strain rate operates) is increasing; see Figure 2.46.

(b) A steady elongation in centimeters per minute, (cm/min) i.e., straight-line behavior in a length/time plot, goes with a diminishing strain rate.

Answer 2.7

No, final length = $(\log_n^{-1} 0.375) \cdot$(initial length)

= 1.45 (initial length),

whichever expression is used for the strain rate.

Answer 2.8

Increase in the region's dimension = 5×1.8 m = 9 m every kilometer; thus strain = 9/1000 and stretch = 1.009.

Answer 2.9

(a) New dimension north-south = 5.60 m,
and east-west = 6.44 m;
new area = 36.064 m².
(b) change of area = 1.064 m².

(c) area strain = 1.064/35 = 0.0304.
(d) Area strain = 0.0304 regardless of the dimensions of the area studied, as long as the strains are uniform throughout the entire region.

Answer 2.10

(a) 1.16   (b) −0.001   (c) − 0.04 per year
(d) −40 per 1000 years. This is exactly the same situation as in part (c). The rate we see *at this instant* is such as would extrapolate to −0.04 in a year or −40 in 1000 years (a mind-boggling concept!), but

we definitely are not making any statement about events occurring during a year or 1000 years; we are simply stating the rate at which we see change occurring now.

Answer 2.11

(a) 9 mm/180 km = 5.10^{-8}, so rate = -5.10^{-8} per year, or −0.05 per million years, or -5.10^{-8} $/3.10^7$ per sec = $(0.17).10^{-10}.17).10^{-14}$ per sec.
(b) The same in magnitude but positive in sign, an elongation rather than a shortening.

(c) (strain rate 5.10^{-8}) × (thickness 40 km) = 2 mm/yr.

It is interesting to compare the answer at (c) with the hypothesis of isostasy and the erosion rate of mountain terrains. If the crust has density 0.9 × (density of mantle) and isostasy is maintained, the base of the crust would *descend* 1.8 mm/yr and initially the land surface would rise 0.2 mm/yr. After a while, sufficient relief would exist for the erosion rate to reach 0.2 mm/yr so that although *particles* continued to rise, the land surface would no longer be rising. In this state, the inward movement of France and Spain is balanced by (i) a rather small-scale return flow of surface sediment and (ii) a larger-scale return flow by some kind of circulation in the mantle.

Answer 2.12

$$\text{Radial strain} = 0.3 \times 0.04 = 0.012. \quad \text{Hence,}$$
$$\text{Volume strain } e_1 + e_2 + e_3 = -0.04 + 0.012 + 0.012 = -0.016.$$
$$F = 0.4 = 1 - 2 \times \text{(spread ratio)}$$

Answer 2.13

Extrusion produces elongation at $0.2 \times 1.6 = 0.32$ per million years. Volume strain rate $= 0.32 - 0.2 - 0.2 = -0.08$ per million years.

Answer 2.14

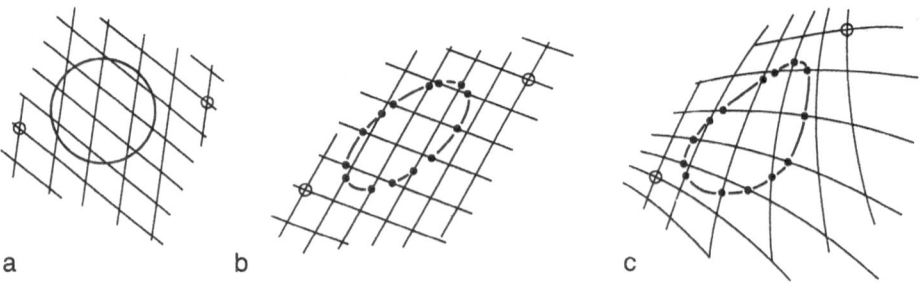

a   b   c

FIGURE 2.47

Long axis: stretch = 1.46, strain = 0.46.
Short axis: stretch = 0.63, strain = −0.37.

Because the strain is homogeneous, the same answers should be reached whatever the position and size of the circle drawn on diagram (a).

Answer 2.15

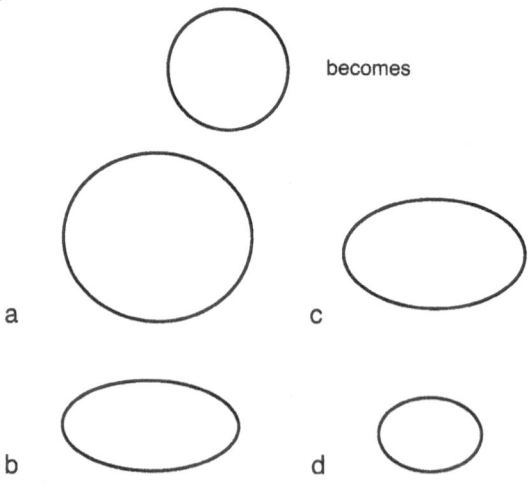

FIGURE 2.48

Answer 2.16

The conclusion to be reached is that the principal directions of strain make 90° at the start of the deformation and make 90° at the end of the deformation.

Answer 2.17

Let R be the radius of the original circle, a and b the axes of the ellipse and r the final length of a radius that made angle θ in the initial state with the line that became the long axis. Then

$$r^2 = a^2\cos^2\theta + b^2\sin^2\theta.$$

For the radius in question, let the stretch be S; then $S = r/R$ and

$$S^2 = (a^2\cos^2\theta + b^2\sin^2\theta)/R^2$$
$$= S_a^2\cos^2\theta + S_b^2\sin^2\theta,$$

where $S_a = a/R$ and $S_b = b/R$

$$= \tfrac{1}{2}(S_a^2 + S_b^2) + \tfrac{1}{2}(S_a^2 - S_b^2)\cos 2\theta.$$

Thus, to get a sine wave (or cosine wave) without assuming the strains to be small, it is necessary to plot (stretch)^2 in a diagram like Figure 2.10(c). But (stretch)$^2 = (1 + e)^2 = 1 + 2e$ if we neglect e^2, giving

$$e = \tfrac{1}{2}(e_a + e_b) + \tfrac{1}{2}(e_a - e_b)\cdot\cos 2\theta \quad (2.23)$$

and, as noted already, results that are approximately true for small strains are exactly true for strain rates.

Answer 2.18

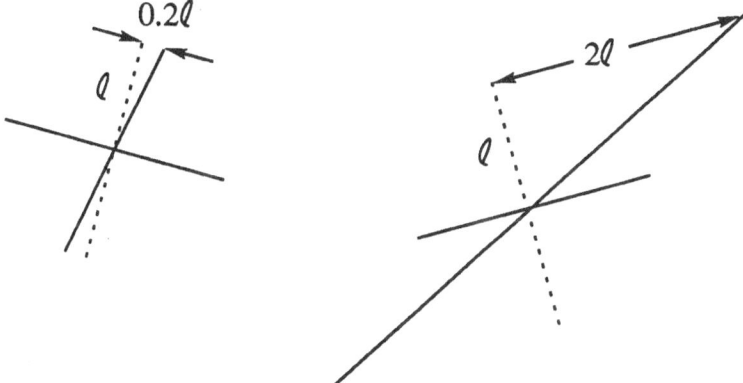

FIGURE 2.49

Answer 2.19

(a) Probably 40° counterclockwise; possibly 400° counterclockwise, or 320° clockwise, etc.
(b) $\phi = 10°$ so shear strain $= \tan 10° = 0.18$.

Answer 2.20

$e_x = 0.46 \qquad e_y = -0.31 \qquad \alpha = 50°$
$\qquad\qquad\qquad\qquad\qquad\qquad \beta = 40°$
$e_x - e_y = 0.77 \qquad \tan \beta = 0.84$

Answer 2.21

No prescribed answer.

Answer 2.22

One has to assume that the fossils were symmetrical at the outset. Then Fossil Z, which is still symmetrical, must have its axis along a principal direction of strain. Then fossil Y is the one to focus on, because its direc-

tions are *symmetrically* disposed with respect to the directions indicated by fossil Z. In the notation of Figure 2.21 and Question 2.20 $\alpha = 53°$, $\beta = 37°$, and maximum shear strain tan $\beta = 0.75$.

Answer 2.23

The distance from the center to the side is 500 mm. Therefore, the displacements of the corner are 500 mm × (0.020) sideways and 500 mm × (0.008) upward. The new position of the diagonal is then a distance Δ from its original position, where $\Delta = 6/\sqrt{2}$ mm, in the neighborhood of the corner, and this separation tapers to zero at the center, a distance $500\sqrt{2}$ mm away. The small angle involved is therefore $\dfrac{6/\sqrt{2}}{500\ \sqrt{2}}$ or 0.006.

The diagonal to the top left corner of the square rotates similarly through an angle of 0.006 in the other direction. Hence the two diagonals, initially at right angles, finally make an angle 0.012 different from a right angle. No lines in the deforming square suffer *more* shear strain than these two, so as needed, the maximum shear strain suffered is 0.012.

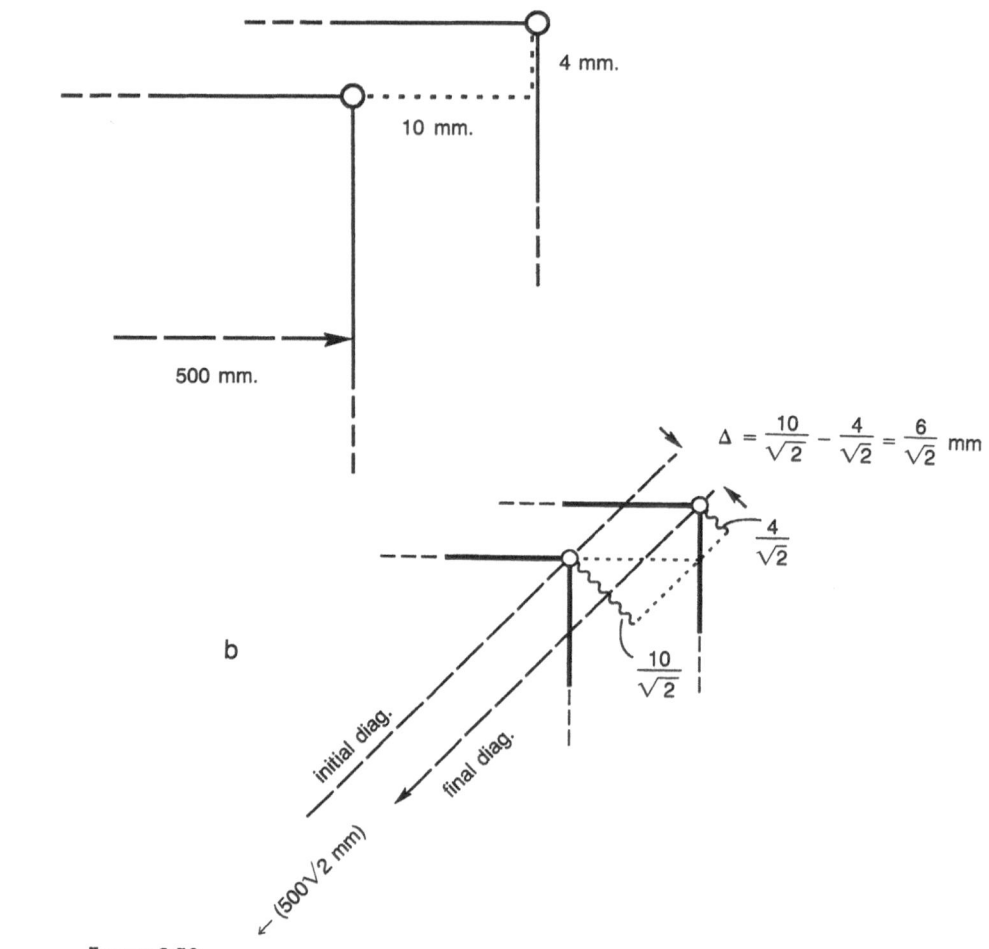

FIGURE 2.50

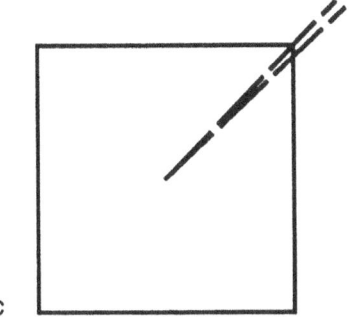

FIGURE 2.50 (*Continued*)

Answer 2.24

(a) Name the lower tip of the marker point O and consider a length L of the marker, with its other end at point P. On account of the shortening, P will move $(L \cos 30)\cdot(0.04)$ parallel to bedding and $(L \sin 30)\cdot(0.02)$ normal to bedding. The first of these displacements has a component normal to L of $[(L \cos 30)(0.04)]\cdot\sin 30$ and the second has a component $[(L \sin 30)(0.02)]\cdot\cos 30$. The total displacement normal to L, which fixes the amount by which L rotates, is the sum of these: $(L \sin 30 \cos 30)(0.04 + 0.02)$. For the amount of rotation, we divide this by the length L of the rotating marker, so that the amount of rotation is $0.06 (\sin 30 \cos 30) = 0.026$.

(b) For the second marker, consider a length M to a tip at point Q. Then on account of shortening, Q moves $(M \sin 30)(0.04)$ parallel to bedding, and the component of this displacement normal to the marker M is $[(M \sin 30)(0.04) \cos 30$. Steps similar to those followed in part (a) lead to the result that the rotation of the second marker is also 0.026.

(c) The total rotation of one marker *with respect to the other* is thus 0.052, an angle of shear; and because this is a small angle, we can also conclude that the amount of shear strain is 0.052.

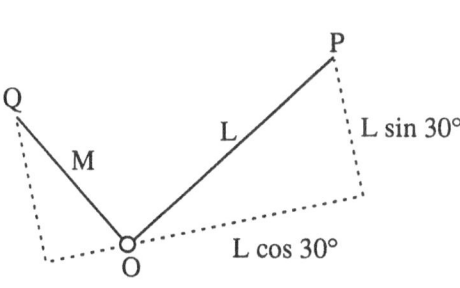

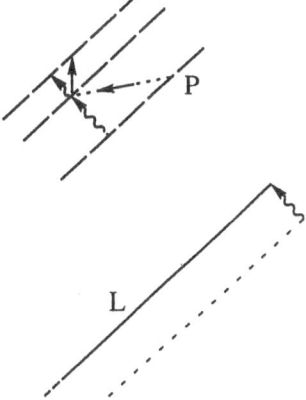

FIGURE 2.51

Answer 2.25

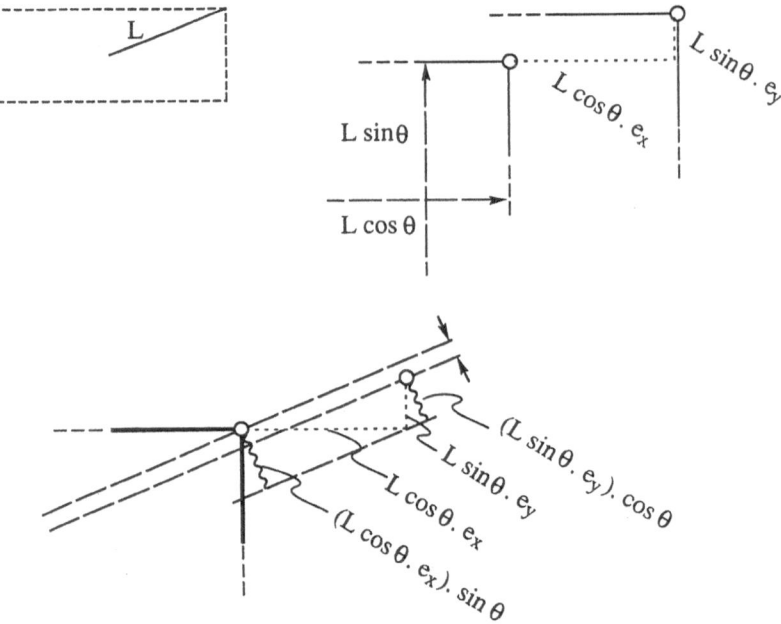

FIGURE 2.52

(a) See Figure 2.52.
Displacements of the tip of the line are $L \cos \theta \cdot e_x$ sideways and $L \sin \theta \cdot e_y$ upward. These displacements have components normal to the direction of the line of magnitude $(L \cos \theta \cdot e_x) \cdot \sin \theta$ and $(L \sin \theta \cdot e_y) \cdot \cos \theta$, for a combined effect $L \sin \theta \cos \theta \cdot (e_x - e_y)$ or $(1/2)(e_x - e_y) L \sin 2\theta$. This separation tapers to zero at the origin O at a distance L away, so that the small angle involved is $(1/2)(e_x - e_y) \sin 2\theta$.

(b) A line at 90° to the line considered in part (a) will make an angle θ with Oy. A similar process of taking components and adding or subtracting shows that this line also rotates through a small angle $(1/2)(e_x - e_y) \sin 2\theta$. Hence either of these rotations is half the total shear strain suffered.

(c) As θ takes all possible values, the maximum value of $1/2 \times$ (shear strain) will be $(1/2)(e_x - e_y)$ and the minimum value will be $-(1/2)(e_x - e_y)$, giving a vertical separation, crest to trough, of $(e_x - e_y)$. For the curve of linear strain, the crest is at e_x and the trough at e_y (or vice versa), so that the vertical separations are the same.

Answer 2.26

Displacements of layer top with respect to layer base are 8 m, 7.2 m, 6.6 m, and 0.8 m. If all these offsets are in the same direction their total is 22.6 m, and the average shear strain for the whole column is 22.6 m/178 m = 0.127.

Answer 2.27

As in Answer 2.3, we need to imagine some definite thickness of succession, such at T m total. Then component parts are $(3/5)\, T$ m of shale and $(2/5)\, T$ m of limestone, and the average shear strain for the whole succession is $(3/5) \cdot (0.32) + (2/5) \cdot (0.12) = 0.24$.

Answer 2.28

For every centimeter of rock thickness measured normal to the slip surfaces, the cumulative slip would be 6×1 mm, or 0.6 cm, and therefore the overall shear strain would be 0.6.

Answer 2.29

(a) 1.3
(b) Using the laminar-flow hypothesis, (present length)/(length at maximum shortness) = 1.7.

The fragments seen might have elongated as well as separating since being at maximum shortness; then estimate (a) would be in error. The deformation may not have been purely a laminar flow; then estimate (b) would be in error. Etc.

Answer 2.30

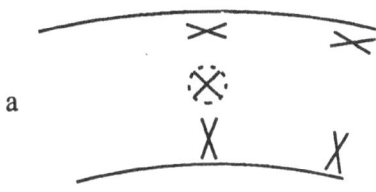
a
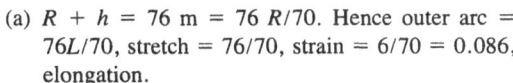

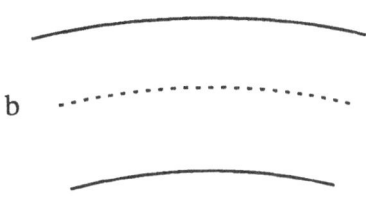
b

FIGURE 2.53

(a) $R + h = 76$ m $= 76\,R/70$. Hence outer arc $= 76L/70$, stretch $= 76/70$, strain $= 6/70 = 0.086$, elongation.
(b) Strain $= -0.086$, shortening.
(c) j has probably elongated, by 0.086, to conserve area.
(d) k has probably shortened, by -0.086.
(e) Maximum shear strain $= e_{max} - e_{min} = 0.172$, the same amount at j and k; zero in the neighborhood of the centerline. For a *circular* arc, the strain at l will resemble the strain at k, and m will resemble j.
(f) The present distance from centerline to outer sur-

face will probably be less than the original h; from the centerline to the inner surface, more than the original h.

If the strain diminishes linearly from margin to centerline, the *average* strain over the outer half of the layer will be $1/2 \times$ (maximum value) $= (1/2)(0.086) = 0.043$; $h = 6$ m and hence shortening $= 0.26$ m. Similarly, for the inner half of the layer, radial elongation $= 0.26$ m. If the radius of the original centerline is indeed 70 m, then the outer surface probably has radius 75.74 m and the inner surface radius 63.74 m—the original centerline is not quite central any longer; see diagram (b) above.

Answer 2.31

(a) 1000 ft $= 300$ m $= 0.3$ km $= g$;
$L = 50$ km;
$R = L^2/8g = 2500/2.4 = 1040$ km.
Half the layer thickness $= 400$ ft $= 120$ m $= 0.12$

km so linear strain, as in A 2.30, $= 0.12/1040 = 0.00012$, shortening. In meters, shortening $= 0.00012 \times 50000 = 6$ m $= 20$ ft.
(b) 0.00012×10 m $= 1.2$ mm   (c) 30 ft, 0.00018

Answer 2.32

(a) In the equation $R = L^2/8g$, we put $R = 6400$ km (the radius of the earth) and $L = 50$ km, giving $g = 50$ m.
(b) The distance just calculated is 1/6 of the distance the center of the limestone layer is depressed. If the two ends of the layer's centerline were rigidly pinned, during the first 50 m of depression, the centerline would shorten; during the next 50 m it would elongate again to its original length; and during the remainder of the depression process, the centerline would elongate beyond its original

length. But the ends of the layer are not in fact rigidly pinned, so these effects need not be considered.

From another point of view, whether the layer is initially straight or slightly curved does not matter; it is the ratio of the depression distance to the length of layer involved that fixes the inward rotation of the radii, as in Figure 2.54. From that point on, one needs to know or assume that at some level the rock does not change length; all strains above or below that level then follow.

Answer 2.33

(a) The circle needed is very like Figure 2.41, with center at (5,0) and radius 3 in length units; each length unit represents a strain rate of 1/yr. The process is an overall swelling, stronger in one direction than in the other; to see such a process one would need a rising bread dough or some similar overall expansion.

(b) The crest-point of the circle represents the behavior of one of the lines of greatest shear strain rate.

(c) The elongation rate of the line of maximum shear strain rate is 5/yr = $(1/2)(\dot{e}_{max} + \dot{e}_{min})$. This fact appears in Figure 2.25, and in the Mohr circle, and in Equation (2.18) at θ = 45° (then cos 2θ = 0).

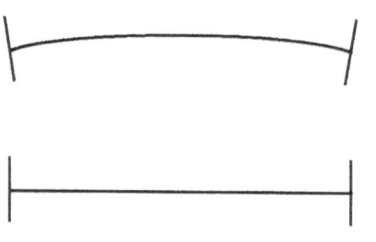

FIGURE 2.54

Answer 2.34

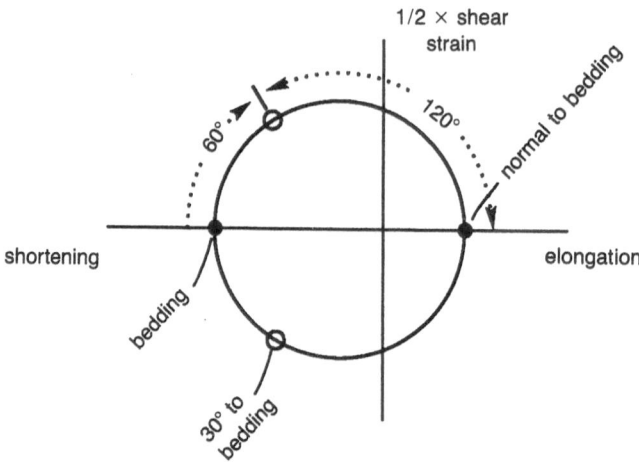

FIGURE 2.55

Either of the open circles in Figure 2.55 serves at part (c). At (e), 1/2 × (shear strain) = 0.026; linear strain = −0.025.

Answer 2.35

(a)	11/1000;	+4,	+3,	−7 thousandths
(b)	4/1000;	+11,	0,	−11
(c)	3/1000;	−1,	−9,	+10
(d)	−7/1000;	+10,	−1,	−9

In (c) the order of the terms is jumbled; obviously, this makes no difference to the correctness of the partition—it is just a custom to list the terms with e_{max} first.

Forces and Stresses

Lofty mountains inspire lofty thoughts: "I will lift up mine eyes unto the hills from whence cometh my help." And structures on outcrop scale, though less majestic, can have great aesthetic appeal. But besides the rich visual texture, a second quality engages the mind: there is a duality, a double vision; is what we see familiar or unfamiliar? On one hand, some aspects of outcrop structures are familiar; "marble cake" looks like marble and conversely, outcrops sometimes look like streaky fluids, pastes, and confections from the kitchen. On the other hand, the time scale is hard to grasp, and the idea of a tombstone gradually changing shape, and yet at any moment during the change being as hard and flinty and apparently resistant to change as it appears today—that idea is less familiar.

Amid these mental exercises, the topics of the present chapter are at the familiar end; the forces and stresses that produce crustal structures are of the same type that exist in walls, chair legs, and butter spreading. As the first section of this chapter shows, the *magnitude* of the compressions may go beyond our direct experience, but the nature of the forces is wholly familiar; one's appreciation of a geological structure can always be enhanced by linking it to an everyday analog. Although the section is short, I hope the reader enjoys the activity practiced, of finding links between geological calculations and daily life.

RELATING FORCES AND PRESSURES TO DAILY EXPERIENCE

A table of units and conversion factors is given inside the back cover.

Question 3.1 Drinking: What force in newtons do you exert lifting a cup of milk or coffee to drink?

Question 3.2 Cycling: A bicycle weighs 30 lb and has tire pressures of 60 psi. What contact area do the wheels have with the ground (a) when the bicycle is standing idle? (b) after a rider mounts who weighs 150 lb?

Question 3.3 Standing: If you rest your hands on a cushion and someone stands on your hands, how does the pressure they exert compare with atmospheric pressure (1 bar)?

Question 3.4 Jacking an automobile: At 30 km depth in the earth, the pressure is about 10 kb or 1000 MPa. What contact area would you need when jacking an automobile to generate such a pressure? (Assume jacking = lifting about 1 ton.)

In the four questions just raised, and constantly throughout the book, the choice exists of expressing forces in newtons or in pounds, kilograms, tons etc. Appendix C discusses the choice. Either policy serves, and readers are encouraged to be familiar with both.

RELATING PRESSURE TO DEPTH

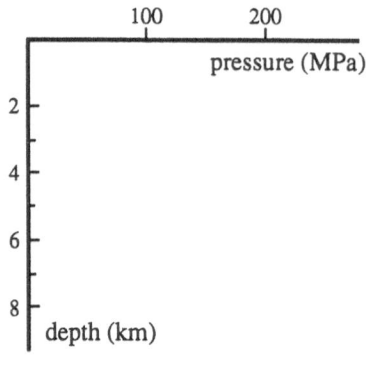

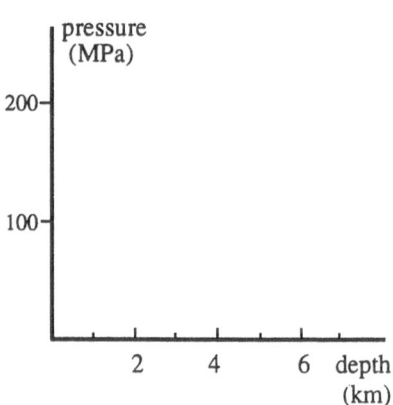

FIGURE 3.1

Queston 3.5 A profile of pressure and depth:

(a) Draw lines in Figure 3.1 to show how pressure increases with depth
  (i) in the ocean. Assume density of water = 1 g/cm³ at all depths, and that 10 m of water gives 1 bar pressure.
  (ii) in a simple uniform earth where all rock has density 3.5 g/cm³.
  (iii) at a site where, from the surface downward, we encounter

2 km of water,
2 km of uncompacted sediment, density 2.0 g/cm³,
2 km of rock, density 3.5 g/cm³,
1 km of pumice (!), density 1.0 g/cm³.

(b) In the second, or right-hand, version of the diagram, how is the slope of each line or line segment related to the density of the rock?

(c) For each segment of the line plotted at (iii), by how much does the pressure change for 1 km change in depth?

The question illustrates the following general idea: over any interval where the density is constant,

$$\underset{\text{MPa}}{\text{change in pressure}} = 10 \times \underset{\text{g/cm}^3}{(\text{density})} \times \underset{\text{km}}{(\text{change in depth})}$$

or

$$\underset{\text{MPa}}{\Delta P} = 10 \underset{\text{g/cm}^3}{D} \cdot \underset{\text{km}}{\Delta h} \qquad (3.1)$$

Even if the density changes continuously with depth, if we focus on a *small* change in depth, *dh*, (maybe a few centimeters) the value the density has right at that level can be used. Thus

$$\underset{\text{MPa}}{dp} = 10 \underset{\text{g/cm}^3}{D} \cdot \underset{\text{km}}{dh} \qquad (3.2)$$

where *D* is now the *local value* of the density, which we can still establish even though we do not wish to assume *D* is constant over any significant interval.

Question 3.6 An increment of pressure: What is the difference in overburden pressure, comparing the top and bottom of a limestone horizon 12.5 m thick if limestone density = 2.4 g/cm³?

Question 3.7 Relation of density to pressure gradient: Why do people drilling for oil add barite to the drilling mud that lubricates the drill stem? Specifically, if water in the formations to be drilled is at a pressure of 0.6 × (rock overburden pressure), what density of drilling mud is needed to prevent the formation water from seeping into the hole? (Assume the rocks drilled are uniform, with density 2.4 g/cm³.)

For more on "overpressure" in water deep in the earth, see Nonuniform Pressure Gradients, page 147.

Instability: Vertical Columns in Parallel but Mismatched

The mechanical behavior of *salt domes* is of interest (and commercially important). Also important is the manner by which water escapes from the lower part of a pile of compacting sediments, for it influences the emplacement of hydrocarbons and ores. Both behaviors can be illuminated by looking at pressure-depth profiles in adjacent vertical columns.

For a salt dome and its surroundings, take densities as shale, 2.7 g/cm³, salt rock 2.3 g/cm³. Then over 3 km or 10,000 ft of vertical depth, the overburden pressure/depth profiles would ideally be as in Figure 3.2. Because of their different gradients, these two cannot match. If they match at the top of the 3-km span, then at the base, p(shale) > p(salt) or if they match at the base, then at the top p(salt) > p(shale); there is also a range of possibilities where the pressures match somewhere within the 3-km interval, leaving p(salt) > p(shale) at the top, and p(shale) > p(salt) at the base. This correlates with the mushroom form salt domes often take, a tendency to spread in the upper parts and to be pinched lower down. Actually, to generate spreading or pinching, we need a mismatch with the *horizontal* pressure, whose relation to the vertical pressure is not always simple; but if we had simple conditions away from the boundary between salt and shale with

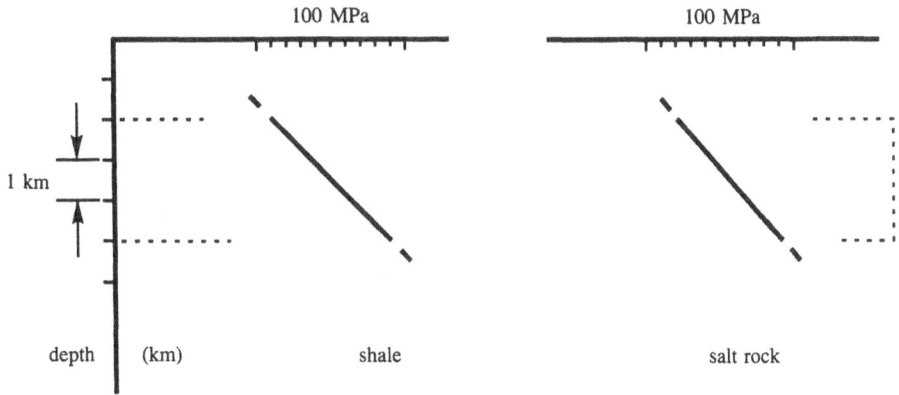

FIGURE 3.2

horizontal pressure equal to vertical pressure, the mushroom shape would be neatly explained.

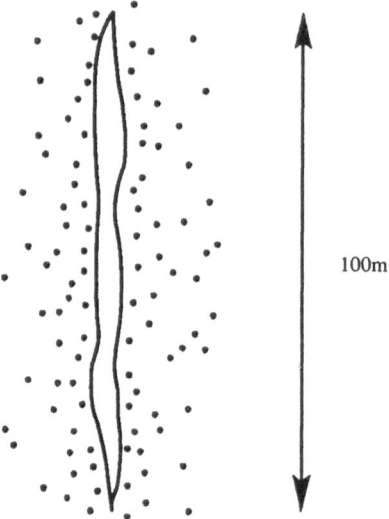

100m

FIGURE 3.3

Question 3.8 Vertical water-filled crack: See Figure 3.3. Continuing from Q 3.7, again assume shale density = 2.7 g/cm; assume water = 1 g/cm, and assume horizontal pressure in shale halfway up the crack = horizontal pressure in water halfway up = vertical pressure in shale halfway up. What is the mismatch in pressure at the top end of the crack, in MPa?

Isostasy

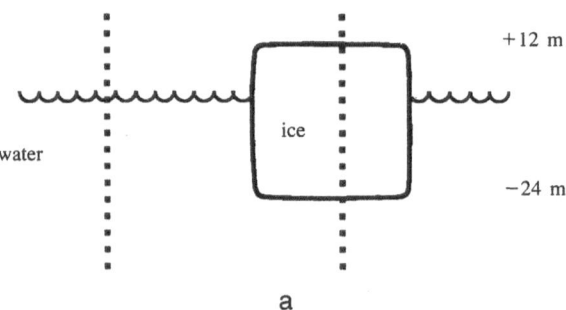

+12 m

ice

water

−24 m

a

FIGURE 3.4A

Question 3.9 Floating ice: Assume, so as to make the diagrams clearer, that the density of ice is 2/3 g/cm³. Draw pressure-depth profiles for the two dotted lines in Figure 3.4(a) (both profiles on one set of axes.)

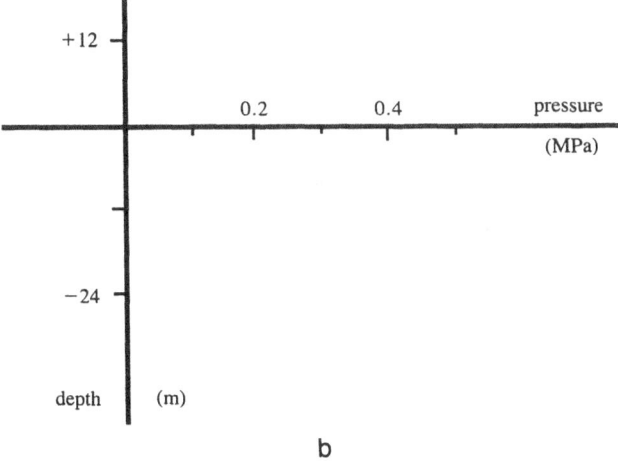

b

FIGURE 3.4B

Besides linking another familiar phenomenon to the idea of a gradient of pressure with depth, Question 3.9 is groundwork for later concepts (Q 3.11 and following).

Question 3.10 Submerged cube: A cube of material that is 20 cm in each dimension is submerged in seawater until its top is 70 cm below the surface, and is held still by magnetic forces. The seawater has a density of 1.1 g/cm³. What water-pressure *force* acts on the cube's top face? on its bottom face? What size bag of potatoes exerts a force equal to the difference between these two?

(The topic is continued at Question 3.11.)

FORCE BALANCE, WEIGHT VERSUS PRESSURE

Forces Vertical

Question 3.11 Submerged cube, continued: Consider the submerged cube in Question 3.10. If, instead of magnetic forces, the cube is held still by a swimmer holding its sides, how much force must the swimmer exert (a) if the cube weighs 5 kg, and (b) if the cube weighs 10 kg? (c) If the cube were held still by a string attached either to an anchor on the seabed or a bridge overhead, instead of by a swimmer, what would be the tensile force in the string at (a), where cube weighs 5 kg, and at (b), where cube weighs 10 kg?

This is our first use of an idea that recurs frequently:

For a sample of material that is not accelerating,

total forces up = total forces down

The topic is continued at Question 3.15, on salt domes, and on page 67, vertical torpedo.

Forces Parallel to a Slope

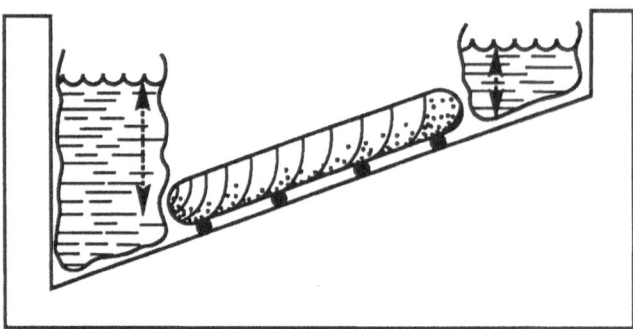

FIGURE 3.5

Question 3.12 Balance of forces, continued: Consider a wood block on rollers on a sloping plane, Figure 3.5. Its weight is a force of 800 g or 8 N (see Appendix C), and the component of this force that acts parallel to the plane is 140 g or 1.4 N. The wood block is controlled by expandable water bags: filling the left bag makes the block roll up, while filling the right bag makes it roll down. The end-area of the block is 12 cm², water depth at left = 14 cm, water depth at right = 3 cm. Will the block roll up the slope or down? (Make a force-balance or imbalance, parallel to the slope; assume the water pressure in each bag is transmitted through the bag to the full 12 cm² of end-area on the block, and that the water is salty, giving 1.1 g/cm³.)

Application to Glaciers   Consider a block on rollers as in Question 3.12 with dimensions l, w and h meters as in Figure 3.6. Let its density be 0.9 t/m³ so that its weight is 0.9 lwh t, where t means the force exerted by a mass of one metric ton. Let its upper end be 1/10 l m higher than its lower end; then the *component* of its weight acting parallel to the slope is 0.09 lwh t. To balance this, the water depth in the left bag must be greater than the depth in the right bag by say, d m. This would give a pressure difference of 1.1 d t/m² and a force difference of 1.1 dwh t. Thus, for forces to be just in balance, we need d $= \frac{0.09}{1.1} l$ m $= 0.082\ l$ m. This is close to the difference in height between the two ends of the block so the water levels in the two bags would be close to level with each other (see Figure 3.7).

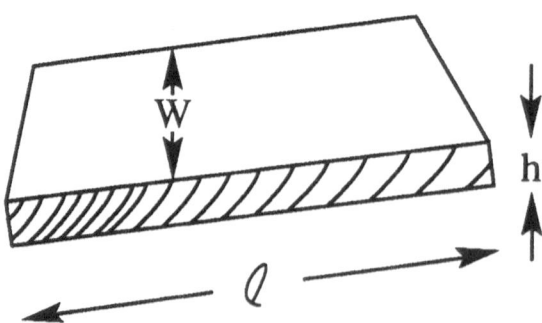

FIGURE 3.6

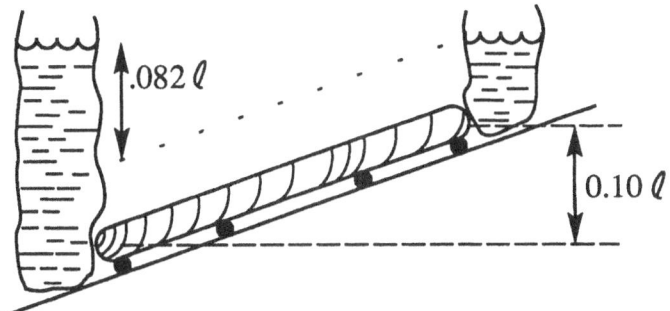

F<small>IGURE</small> 3.7

An important idea emerges if we switch from water with density 1.1 g/cm³ to special water with density 0.9 g/cm³. Then, for balance, the depth-difference in the two bags needs to *exactly* equal the height difference of the two ends, and the water levels in the two bags need to be exactly level with each other. Or, instead of special water, we could fill the bags with solid ice at density 0.9 g/cm³, and the forces would still balance as long as the bags or columns of ice continued to have the ability to exert slope-parallel pressures equal to their (depth × density) pressures (hydrostatic or lithostatic or whatever). But if the left ice-column reached higher, the block would tend to move up-slope, and if the right ice-column reached higher, the block would tend to move down-slope. This can be restated: the direction of the block's movement, to left or to right, is always from the higher ice-column toward the lower ice-column. That is, it is the direction of the *surface* topography's downward gradient that determines the direction the ice moves, regardless of the basal topography, (as long as the latter is not too peculiar or extreme).

The same idea can be applied to thrust sheets. It is slightly erroneous because columns of rock normally do not exert sideways pressures precisely equal to their nominal (depth × density) pressures, and overly simple because the central block is apt to deform, rather than retain a constant shape while moving to and fro. But it is a powerful, approximately correct idea, that has been used to aid the search for oil in midwestern North America.

The topic is continued under headings *Change of Stress from Point to Point*, page 77, and *Inhomogeneous Rock and Thrust Sheets*, page 175.

Forces Vertical, Pressures Partitioned over Two Areas

Question 3.13 Balance of forces—a chair-borne emperor: Two bearers carry an emperor whose weight, including ornaments, is 102 kg. The front bearer's shoulder pads have a total contact area of 25 cm² and an average pressure of 2 kg/cm². The back bearer's shoulder pads have a total area of only 13 cm²: What is the average pressure in these? (Here "kg" means the force exerted by a mass of 1 kg, as discussed before.)

Question 3.14 Balance of forces—house walls: The upper part of a two-story house weighs 24 t and is supported on exterior walls (total cross-section area = 6 m²) and interior walls (total cross-section area = 3 m²). If the stress or pressure in the exterior walls is 3 t/m², what is the pressure in the interior walls?

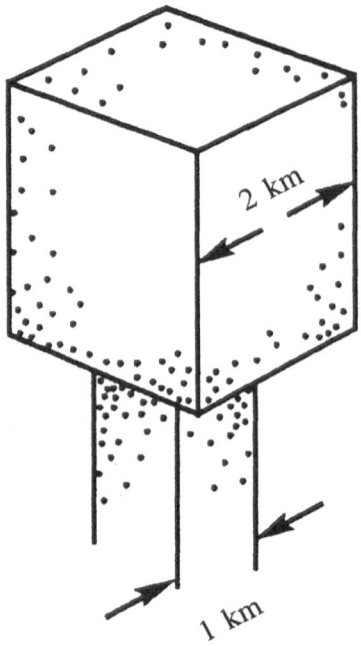

FIGURE 3.8

Question 3.15 Balance of forces—salt dome: Some salt domes are mushroom-shaped, with a swollen upper part and attenuated stem (compare with other buoyant deformable bodies). Consider the upper part of a salt dome that approximates a cube measuring 2 km each way with its top surface at ground level (Figure 3.8); it is surrounded by shale except for its stem, which is 1 km² in cross section. Salt density = 2.3 g/cm³ = 2.3 t/m³; shale density = 2.6 t/m³; assume that at 2 km depth where shale meets the underside of the salt overhang, the shale exerts as much *upward* pressure as the regular shale overburden's downward pressure at other points 2 km deep. Does the stem of the salt push upward or pull downward on the cubic head of the dome? with how much pressure or tensile stress?

Question 3.16 Salt dome, continued: Repeat Question 3.15 using the same data, but now assume that the stem cross-section in only (1/2 km)² or 1/4 km² so that the upward shale pressure operates on 3 3/4 km².

Weighted Mean Pressure

Question 3.17 Two areas with different pressures— A thin horizontal coal seam was worked at 320 m below ground where the overburden pressure was 800 t/m², (*Check:* Is this the right order of magnitude, from your knowledge of rock density? Three-quarters of the seam was removed, leaving pillars for support occupying one-quarter of the horizontal area. The mine is now abandoned and flooded, with water pressure at the level of the seam = 300 t/m². What is the average pressure in the pillars?

Note: to use TFU = TFD, we need some areas. You may choose to consider a representative horizontal area at the depth of the seam of, say, 50 m² or 1 km² or A m².

Question 3.18 Areas with different pressures, continued: A sample of pumice is 80 percent holes, 20 percent rock; this is true by volume, and also any representative cross section would show *areas* in these ratios.

(a) If the average total pressure on the pumice is 24 MPa, and the holes contain a fluid at 12 MPa, what is the average pressure in the rock portion of the pumice?

(b) If the average pressure in the rock portion is 32 MPa, what pressure must the pore-fluid be carrying? (Total pressure is still 24 MPa.)
(c) Repeat part (a), but now consider a sandstone that is 75 percent rock and only 25 percent holes. What is the average pressure in the rock portion of the sandstone?
(d) What state develops in the pumice if the fluid pressure rises to 32 MPa?

This example leads into a large field of possible problems. Processes in the earth are much affected by fluids: besides petroleum in sandstones, ore veins, tension gashes and many other fractures are consequences of fluid activity. In volcanoes, fluid pressure is sometimes sufficient to blow the rock apart, and metamorphic minerals liberate fluids with rise of temperature despite ambient compression of say 500 MPa, 5 kb, 50000 t/m², 5 t/cm² (cf. the automobile jack, Question 3.4). But however unfamiliar the idea of pervasive fluids at such high pressures, one can always rely with confidence on the basic rule "total forces one way = total forces the opposite way." It comes up in many disguises.

Question 3.19 Areas with different pressures, continued: A sedimentary succession consists of horizontal alternating beds of shale 12 cm thick and limestone 8 cm thick. The whole succession suffers an average horizontal compressive stress of 54 MPa. (a) If the average horizontal compressive stress in the shale is 60 MPa, what is the average horizontal compressive stress in the limestone? (b) If the average horizontal compressive stress in the shale is 95 MPa, what is the average horizontal compressive stress in the limestone?

(Compare with the coal seam (Q 3.17) and pumice (Q 3.18).)

Generalization

The ideas underlying Questions 3.17 through 3.19 are conveniently expressed if we use the idea of an **area fraction.** In the coal seam,
the area fractions were   1/4, 3/4, or 0.25, 0.75;
in the pumice, they were   20%, 80% or 0.20, 0.80;
in the sedimentary succession 8/20, 12/20, or 0.40, 0.60;
and for any regular partition over two components, we can find two area fractions that add to 1.0. If two such area fractions are a_1 and a_2 (where $a_1 + a_2 = 1$) with pressures p_1 and p_2 respectively, the average pressure $= a_1p_1 + a_2p_2$. If we write p for the average pressure, then

$$p = a_1p_1 + a_2p_2 \tag{3.3}$$

and this *looks* like a relation among pressures, but it is not: we have only to reinsert a representative area

$$\underset{\substack{\text{total} \\ \text{force}}}{pA} = \underset{\substack{\text{force in} \\ \text{component 1}}}{(a_1A) \cdot p_1} + \underset{\substack{\text{force in} \\ \text{component 2}}}{(a_2A) \cdot p_2} \tag{3.4}$$

to be reminded that, although a_1 and a_2 are not areas but only ratios, still the equation $p = a_1p_1 + a_2p_2$ is a *force* balance, slightly disguised.

Weighted Mean   The pressure p, called here the "average pressure" in a colloquial way, might also be called the "mean pressure.' But it is a *weighted* mean because the two component pressures p_1 and p_2 are weighted by their relative importance—that is, by the fraction they act on of the total area pressed.

The density of a mixture such as sand and sugar is another example of a weighted mean: $d = v_1d_1 + v_2d_2$ (assuming no porosity). Questions 2.3 and 2.27 dealt with weighted mean strains.

Question 3.20 Salt dome, continued: Repeat Question 3.16 using the same data, but now assume that the top of the salt cube is at depth 3 km and the base is at depth 5 km. What is the force and the stress in the stem?

FORCE BALANCE, SHEAR STRESS VERSUS WEIGHT

If we consider a small portion of surface inside a continuum, the material on one side pushes or pulls the material on the other side with a tangential drag as well as a normal pressure, and the definition

$$\textbf{shear stress} = \text{(tangential force)/(area)}$$

parallels the definition

$$\textbf{normal stress} = \text{(normal force)/(area)}.$$

The rule, "where materials are not accelerating, total forces one way equal total forces the other way" can involve shear stresses and shear forces in the same way as normal stresses and normal forces.

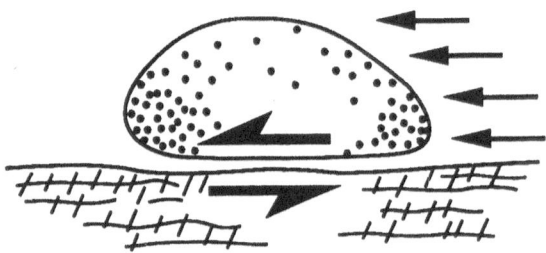

FIGURE 3.9

Question 3.21 Balance of tangential forces: See Figure 3.9; moving ice exerts a force on a boulder of 1320 t (using 1 t as a unit of *force;* see Appendix C). If the contact area between the boulder and the glacial pavement is 1.1 m², what average tangential stress is produced? Figure a value in t/m² and in bars and MPa. Compare with the amount by which the tangential stress on a fault drops when the fault slips in an earthquake, which is often 100 to 500 bars. The magnitudes should be comparable, because creating glacial rock-flour and creating fault-gouge are similar processes.

Force Balance, One Shear Stress and Weight

Question 3.22 Balance of tangential forces, continued: A boulder of weight 1320 t rests on a slope of 25°. What is the average tangential stress at the surface where the boulder meets the slope, if the contact area is 1.1 m²?

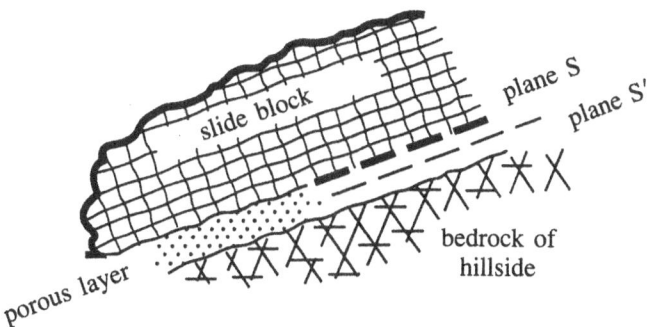

FIGURE 3.10

Question 3.23 Stability of hill slopes: Consider a porous rock where the porosity = 0.2. Suppose this porous material supports a block that is in danger of sliding downhill, as in Figure 3.10, and that the tangential stress parallel to plane S is 4 MPa (40 bars) due to the weight of the slide block.

(a) What is the average tangential stress in the *rock part* of the porous layer parallel to plane S'? (Assume that the tangential stress in the pore fluid is zero).

Whether the rock of the porous layer fails or withstands this shear stress depends on the compression normal to plane S', as shown in Figure 3.11. Suppose compression normal to S due to the weight of the slide block is 30 MPa (300 bars).

(b) What is the compressive stress in the *rock part* of the porous layer if the pore-fluid pressure is negligible? if the pore-fluid pressure is 20 MPa?

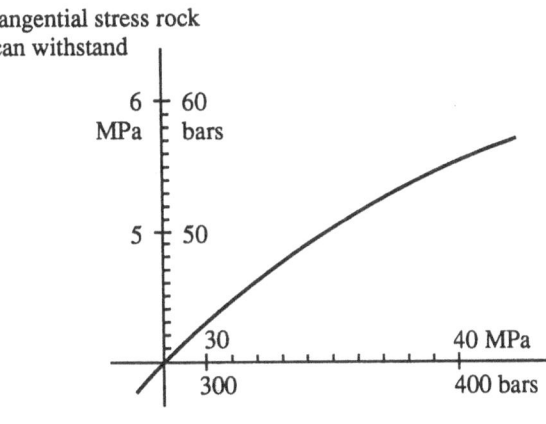

FIGURE 3.11

Force Balance, Two Shear Stresses and Weight

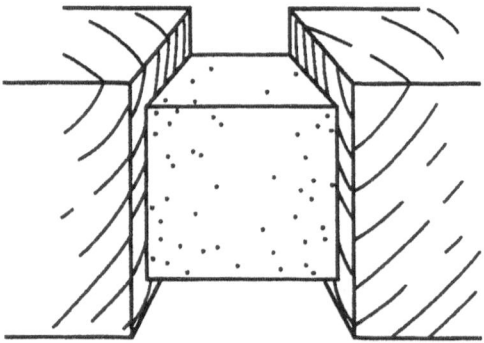

FIGURE 3.12

Question 3.24 Block supported in a vise: See Figure 3.12. If the block shown is an 8-cm cube, of mass 2560 g, what is the average shear stress on the two side faces (whose total area is 128 cm²)? (Use the weight of 1 g as a unit of force, or assume 1 N = the weight of 100 g.)

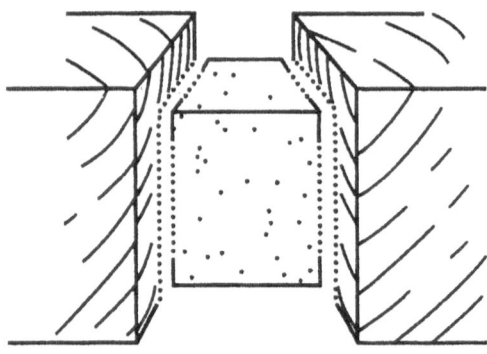

FIGURE 3.13

Question 3.25 Block supported in a vise, interior surfaces: If we consider just the middle 6 cm of the block, with weight 1920 g, as in Figure 3.13, its weight is balanced by upward shear stresses on the vertical surfaces 6 cm apart (total area = 128 cm² as before).

(a) What is the average shear stress on these two surfaces?

(b) What is the average shear stress on two parallel surfaces bounding just the middle 4 cm of the block?

Question 3.26 Block in vise, continued: Make a plot of the answers to Questions 3.24 and 3.25, parts (a) and (b), and generalize them to a continuous graph.

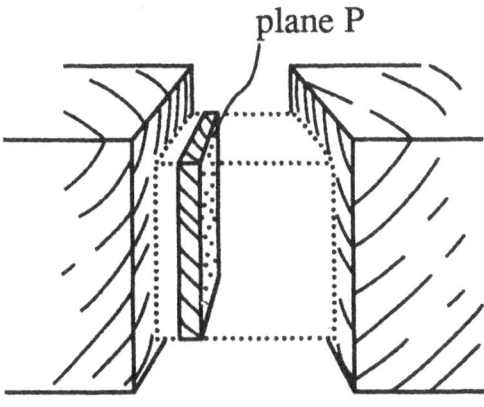

plane P

FIGURE 3.14

Question 3.27 Block in vise, completion: Consider just that portion of the cube that is between 2 cm and 3 cm from the middle, as in Figure 3.14. What combination of shear stresses supports its weight and keeps it from accelerating?

(The topic of this section is continued in Question 3.32, on mountain building.)

FORCE BALANCE, SHEAR STRESS VERSUS PRESSURE

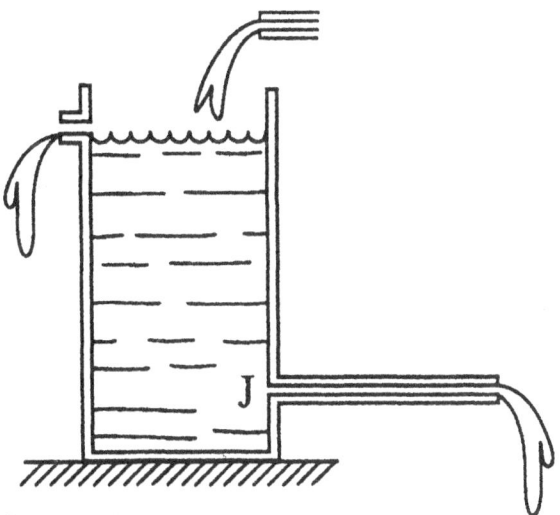

FIGURE 3.15

Question 3.28 Steady flow in a horizontal pipe: The flow system is shown in Figure 3.15. The radius of the pipe is 0.8 cm, its length is 5 m, and the water level is 6 m above the outlet, so that the pressure at J is 0.6 bar or 0.06 MPa. With what tangential stress does the flowing water drag on the pipe's inner surface? (1 MPa $-$ cm^2 is a permissible unit of force for a force balance.)

Note: The answer does not depend on the viscosity of the fluid; for a tank of fuel oil or a tank of honey, the balance of forces would be the same (except that the pressure at J would be a little different on account of different density). But the *response* to the stresses (i.e., the flow rate) could be very different.

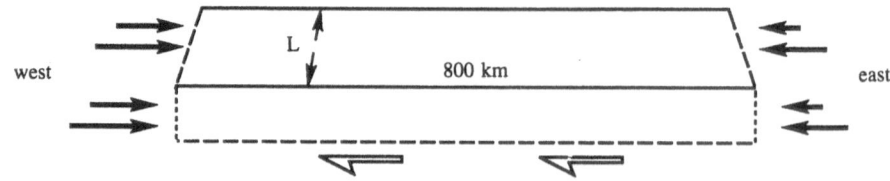

FIGURE 3.16

Question 3.29 Force balance on a lithosphere plate: A portion of ocean-floor lithosphere, Figure 3.16, measures 800 km east-west, L km north-south, and 80 km thick. The horizontal compressive stress on its western edge is, on average, 60 MPa larger than the compressive stress on its eastern edge. If the extra 60 MPa is wholly balanced by shear stress on the base of the slab, what is the average magnitude of this shear stress?

(The topic is continued on page 115 under the heading Is Plate Tectonics True?)

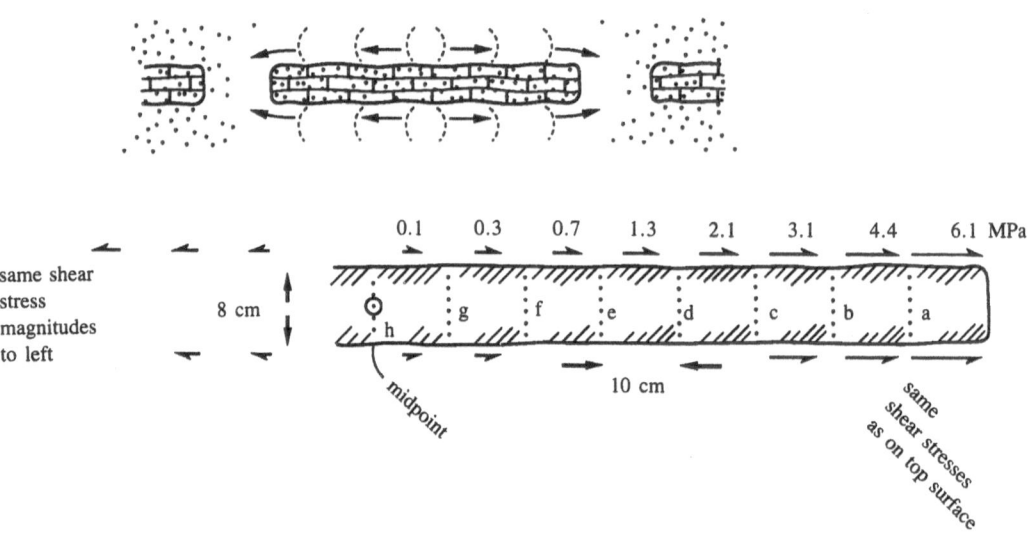

FIGURE 3.17

Question 3.30 Boudinage: A sedimentary succession consists of horizontal alternating beds of shale 12 cm thick and limestone 8 cm thick, as in Q 3.19. Under a heavy load, the succession is spreading sideways and thinning from top to bottom, and the limestone layers have already parted into boudins. One boudin is shown in Figure 3.17, which shows the average shear-stress magnitudes exerted as the shale spreads freely sideways, dragging on the limestone as it goes, for eight segments each 10 cm long. If the compressive stress on the end of the boudin is 23 MPa (average), what is the average stress on each of the planes a, b, c etc?

Suggestion: Write a force balance, considering W cm of rock in the direction perpendicular to the plane of the diagram.

FORCE BALANCE: SHEAR STRESS, WEIGHT, AND PRESSURE

Revisit Q 3.24 and note that an idea of great importance emerges if we ask, What is the pressure on the undersurface of the cube? It is just atmospheric pressure and is *not* the pressure of 8 cm of overburden. The idea that, inside the earth, pressure on a horizontal surface equals density times gravity times depth ($\rho g h$) is correct *only where vertical shear stresses are absent* (or is in error by an amount equal to whatever vertical shear force is present).

Over much of the earth's interior, pressures are probably close to lithostatic; vertical shear stresses are linked with vertical relative motions—basins sinking, domes rising, etc.—and many regions are free of such effects. But the proper form of force balance, TFU = TFD should include shear effects, thus:

total force up = $p_{bottom} \times$ (area)$_{bottom}$ + any upward shear force;
total force down = $p_{top} \times$ (area)$_{top}$ + weight + any downward shear force.

Question 3.31 Force balance: shear stress, weight, and pressure: Revisit Question 3.11. If the contact area of the swimmer's hands on the cube is 10 cm², what average shear stress operates to hold the cube stationary in each of the two situations?

Force Balance with Moving Body

Concerning the submerged cube held stationary by a swimmer, what would happen if the swimmer let go? For a short time, the cube would accelerate, but we can envisage a terminal velocity and a steady motion where, as in other calculations, we assume acceleration equals 0. Then vertical forces must once again balance, for the moving cube; can we write numbers for the parts that add up and balance out? As a matter of fact, a cube is not streamlined and the water-flow around it is vastly complicated, so we consider instead a streamlined object that sinks or floats—a vertical torpedo.

Vertical Torpedo, or Submerged Vertical Pencil   A pencil has dimensions as shown in Figure 3.18 and is held under water of density 1.1 g/cm² by a swimmer:

water-pressure force on top end = $p_{top} \times$ 64 mm²;
water-pressure force on bottom end = $p_{bottom} \times$ 64 mm²;
difference = 64 mm² $\times \Delta p$ = 64 mm² $\times$ 22 g/cm²
$\qquad$ = 14.08 g or 0.14 N.

so the force balance is

downward force by swimmer of 2.60 g + weight of 11.48 g
= net upward force due to water pressure of 14.08 g.

Now the swimmer sharpens the pencil at both ends; this has negligible effects on the force balance and ensures that, even when the pencil torpedoes upward, the end-pressure difference stays close to 14.08 g (no build-up of pressure ahead of pencil's front end). When the swimmer releases the pencil and it achieves a steady upward velocity, what keeps the forces balanced? A tangential drag or shear stress on the vertical sides (area 4 $\times$ 1600 mm² = 64 cm²) of magnitude 2.60/64 or 0.04 g/cm².

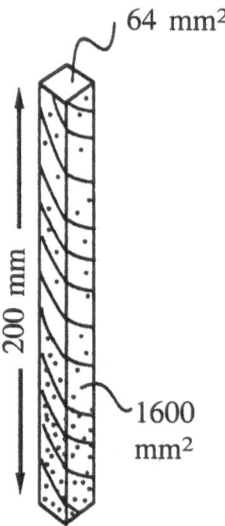

weight 11.48 g

FIGURE 3.18

The purpose of the pencil calculation is to shed light on the presence of vertical shear stresses in the earth; a rising salt dome, for example, has shear stresses on vertical surfaces around it, that slow down its rise. Let us look again at the cube-shaped salt dome in Q 3.20 surrounded by shale, as shown in Figure 3.19. One possible force balance was calculated by assuming that no vertical forces act on the sides:

TFD = shale pressure force on top + weight of salt;
TFU = shale pressure force on bottom (3 3/4 km²)
     + salt pressure force on bottom (1/4 km²)

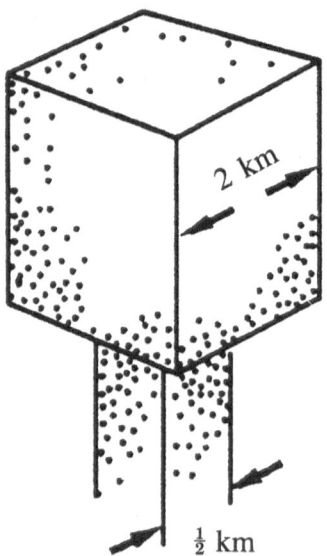

FIGURE 3.19

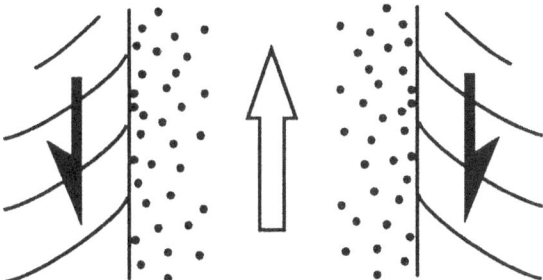

FIGURE 3.20

As a matter of fact, the upward salt pressure is not likely to be as small as in Answer 3.20; it is likely to be closer to the shale pressure. So a more realistic force balance would show a larger TFU, and TFD would include (shale pressure force on top) + (weight) + a third term due to vertical downward tangential stress on the sides as the salt cube rises, deforming the shale (see Figure 3.20).

It is interesting to compare the vertical torpedo or pencil with the fluid-filled fissure (see Question 3.8). In both cases, gravity plus a density contrast drives upward motion; the main difference is that, in the fissure, the highly fluid part is the part that rises, in almost rigid surroundings, whereas with the pencil, the highly fluid part is the surroundings. The fluid flow patterns at the interface can be contrasted, as in Figure 3.21, but the contribution to the force balance from vertical shear stresses at the interface is much the same.

Pyrenees Mountains could perhaps be formed if two rigid portions of the lithosphere came together like the jaws of a vise, trapping a more deformable portion of the lithosphere between them. All real mountains have a history more complicated than this, but the Pyrenees can be used as an example as if that had been their history, with France and Spain as the jaws, movements north-south, and the Pyrenees forming east-west (Figure 2.4). Then we can introduce the mechanics by considering an idealized cross-section, as was done in Question 2.11.

In the situation envisaged, France and Spain move as rigid blocks, and a long, narrow strip of deformable material between them is shortened north-south horizontally and

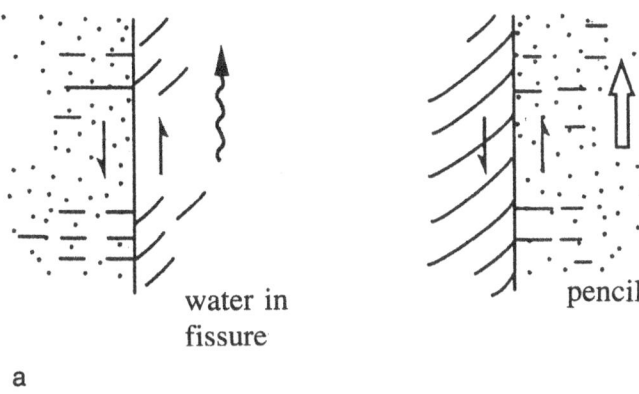

water in
fissure

a   b

FIGURE 3.21

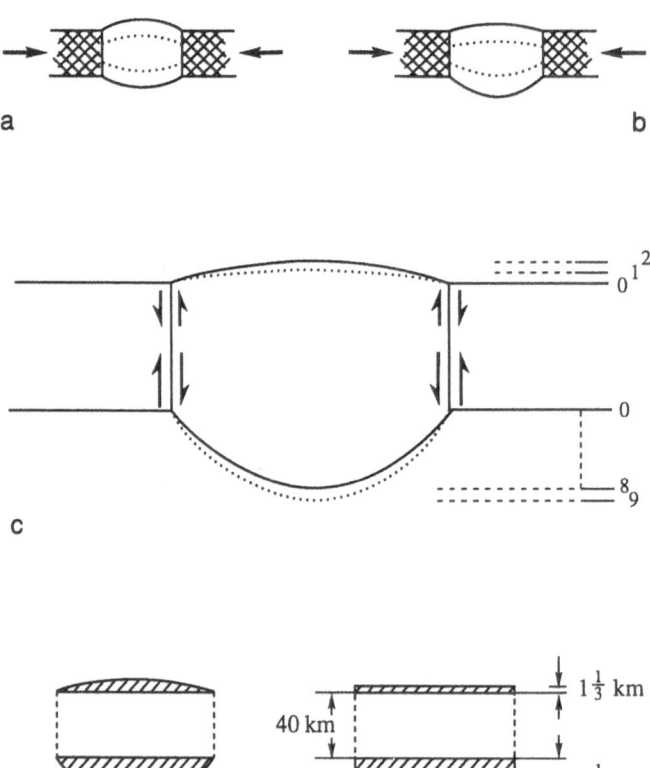

FIGURE 3.22

thickened vertically. If there were no forces of gravity, as much material would bulge upward as bulges downward, as in Figure 3.22(a); but in fact there is gravity and the bulging is less symmetrical, as in Figure 3.22(b). How asymmetrical is this effect?

In Answer 2.11, the idea of isostasy was used. If the density ratio of crust to mantle is 0.9 (e.g., 2800 kg/m³ in the crust and 3100 kg/m³ in the mantle) a bulge upward of 100 m that displaces air is balanced by a bulge downward of 900 m that displaces mantle material, and if all the materials are standing still, this is the state of affairs one can expect. But during a period when the materials are in motion the force balance is not so simple.

The point to notice is that the barreling effect involves shearing. The two dotted lines in Figure 3.22(a) were originally horizontal, as were the top and bottom surfaces of the deforming strip; that is to say, originally the dotted lines made 90° with the vertical walls. But as deformation proceeds, the angle changes; at the upper left wall the motion is just as in Figure 3.21(a), and the deforming mass exerts an upward shear stress on the wall. Similarly at the lower left, it exerts a downward shear stress. If bulging were symmetrical, the net shear stress would be zero, but if bulging is asymmetrical, as in Figure 3.22(b), the upward and downward effects do not balance. The net effect is that the central strip exerts downward shear forces on the walls, and the walls exert upward shear forces on the strip.

Question 3.32 Vertical force balance in mountain building: Suppose that, while France and Spain are approaching, the central strip bulges upward by 2 km and downward 8 km (instead of the isostatic condition, upward 1 km and downward 9 km). See Figure 3.22(c), but assume that for the purpose of balancing forces, the curving extrusions can be replaced by rectangular slabs two-thirds as thick, as in diagram (d).

(a) What total shear force must act vertically upward at the sides, assuming the Pyrenees are L km long from sea to sea? (*Suggestion:* Calculate how much of the upper 1 1/3 km can be supported isostatically by the 5 1/3 km root; whatever remains of the 1 1/3 km has to be supported by shear forces. Assume density = 2800 kg/m³.)

(b) What *average* shear stress at the walls would provide the needed force? (Total area of wall = 80 L km².)

(c) What fraction is this of the vertical compressive stress under 40 km of the earth's crust?

The ideas embodied in Answer 3.32 are not wholly unrealistic: shear stresses of such magnitudes do occur in the earth. Also the height the Pyrenees attained during formation probably was greater than their ultimate isostatic height. In the process described, they bulge upward to an excessive extent and later, as the driving inward motion of the jaws tapers off, the mountains settle back closer to an isostatic state. (There are effects of erosion as well, as discussed at Answer 2.11.)

On the other hand, treating the crust in isolation is not realistic; it is crust and upper mantle together that make up the lithosphere, as in Figure 2.4, which shows two slabs each about 100 km thick approaching each other. Adapting Question 3.32 to treat the upper mantle as well as the crust is not straightforward because the crust and upper mantle are different mechanically and the behavior of each is interfered with by the other.

SUMMARY AND GENERALIZATIONS

Any direction can be chosen for a force balance. As long as, parallel to this direction, a body's acceleration is zero, then total forces one way equal total forces the other way. The simplest shapes to treat are orthogonal, with faces either parallel or perpendicular to the direction of interest. Then the total forces along direction d are, as in Figure 3.23,

$$W \cdot \cos \alpha + p_1 \cdot bc - p_2 \cdot bc + t_1 \cdot ac - t_3 \cdot ac + t_2 \cdot ab - t_4 \cdot ab, \qquad (3.5)$$
$$\text{due to weight}$$

and when the forces balance, this total equals 0. Many different-looking equations are restatements of this basic idea.

1.   Divide by the volume of the block, abc:

$$\rho g \cdot \cos \alpha + \frac{p_1 - p_2}{a} + \frac{t_1 - t_3}{b} + \frac{t_2 - t_4}{c} = 0.$$

2.   If a, b, and c are small, (1) can be rewritten:

$$\rho g \cdot \cos \alpha + \frac{\partial p}{\partial x} + \frac{\partial t}{\partial y} + \frac{\partial t}{\partial z} = 0.$$

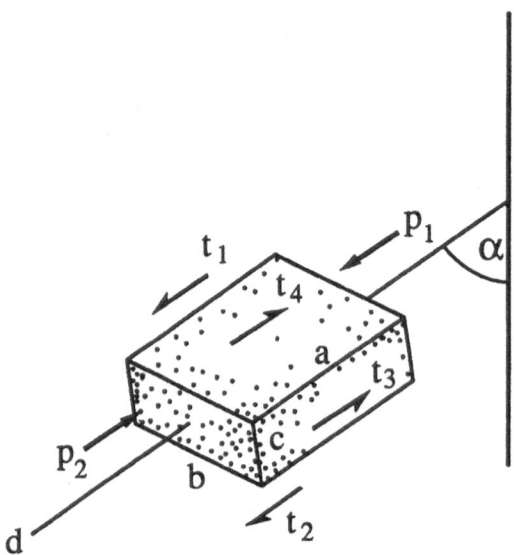

FIGURE 3.23

Special cases:

(a) $\alpha = 0$, vertical force balance:

(i)   if all t's $= 0$, $\dfrac{\partial p}{\partial x} = -\rho g$

as in Question 3.5, pressure gradients with depth;

(ii)   if all p's $= 0$ and $t_2 = t_4 = 0$, $\dfrac{\partial t}{\partial y} = -\rho g$

as in Question 3.24, block in vise.

(b) $\alpha = 90°$, horizontal force balance (or rock's own weight is negligible in comparison with other forces acting on it):

$$\frac{\partial p}{\partial x} = - \left(\frac{\partial t}{\partial y} + \frac{\partial t}{\partial z} \right),$$

rather like Question 3.28, flow in a horizontal pipe;
or, in two dimensions,

$$\frac{\partial p}{\partial x} = - \frac{\partial t}{\partial y}$$

as in Question 3.29, lithosphere slab, and Question 3.30, boudinage.

3.   Use x_1 x_2 x_3 for axes instead of x y z, use σ_{11}, σ_{22}, etc. as shown in Figure 3.24, and use W_1 W_2 W_3 for the components of the weight of unit volume of material along directions x_1 x_2 x_3: then

$$W_1 + \frac{\partial \sigma_{11}}{\partial x_1} + \frac{\partial \sigma_{21}}{\partial x_2} + \frac{\partial \sigma_{31}}{\partial x_3} = 0$$

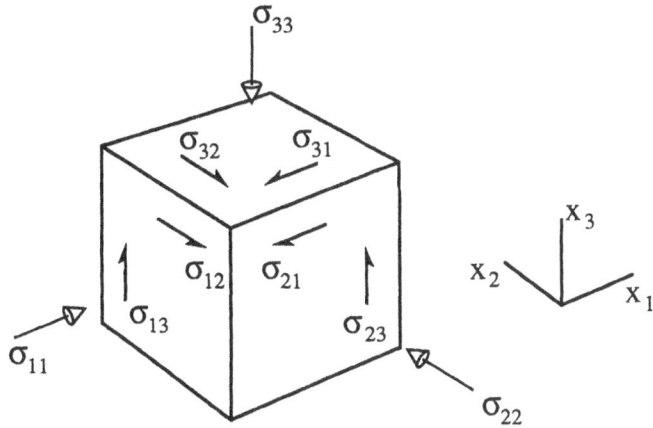

FIGURE 3.24

also

$$W_2 + \frac{\partial \sigma_{12}}{\partial x_1} + \frac{\partial \sigma_{22}}{\partial x_2} + \frac{\partial \sigma_{32}}{\partial x_3} = 0$$

and

$$W_3 + \frac{\partial \sigma_{13}}{\partial x_1} + \frac{\partial \sigma_{23}}{\partial x_2} + \frac{\partial \sigma_{33}}{\partial x_3} = 0$$

4. Let i stand for 1, 2, or 3; then

$$W_i + \frac{\partial \sigma_{1i}}{\partial x_1} + \frac{\partial \sigma_{2i}}{\partial x_2} + \frac{\partial \sigma_{3i}}{\partial x_3} = 0,$$

and let $\dfrac{\partial \sigma_{ji}}{\partial x_j}$ stand for the sum of three terms $\dfrac{\partial \sigma_{1i}}{\partial x_1} + \dfrac{\partial \sigma_{2i}}{\partial x_2} + \dfrac{\partial \sigma_{3i}}{\partial x_3}$. Then

$$W_i + \frac{\partial \sigma_{ji}}{\partial x_j} = 0.$$

5. The three equations in paragraph 3 can be rewritten in **vector-matrix form:**

$$W_1 = - \left(\frac{\partial \sigma_{11}}{\partial x_1} + \frac{\partial \sigma_{21}}{\partial x_2} + \ldots \right)$$

$$W_2 = - \left(\frac{\partial \sigma_{12}}{\partial x_1} + \ldots + \ldots \right)$$

$$W_3 = - \left(\frac{\partial \sigma_{13}}{\partial x_1} + \ldots + \ldots \right)$$

and the three W's are the components of the total weight, which is a vector; so the equations can be written in vector-matrix form

$$\begin{bmatrix} W_1 \\ W_2 \\ W_3 \end{bmatrix} = - \begin{bmatrix} \dfrac{\partial}{\partial x_1} & \dfrac{\partial}{\partial x_2} & \dfrac{\partial}{\partial x_3} \end{bmatrix} \begin{bmatrix} \sigma_{11} & \sigma_{12} & \sigma_{13} \\ \sigma_{21} & \sigma_{22} & \sigma_{23} \\ \sigma_{31} & \sigma_{32} & \sigma_{33} \end{bmatrix}$$

or $\mathbf{W} = -\nabla \sigma.$

This last statement condenses the 12 terms and three equations from paragraph 3 into very few symbols, but the thoughts condensed are all encompassed by:

(i)   force = stress × area

(ii)   where acceleration = 0, total forces one way equal total forces the other way.

Written out in full, the equations look clumsy, and written in condensed form, they look obscure but, either way, the underlying ideas are rather simple:

> Forces balance.
> Orthogonal directions are independent.

The question now arises: How far can the ideas presented take us? Turning to real rocks and to outcrops, surely we soon come to situations that are not so geometrically tidy, where the planes we need to consider are not orthogonal. This is certainly true, and Chapter 4, where oblique planes are discussed, contains essential material. But before we turn to oblique planes, two more topics can be explored using orthogonal planes or brick-shaped representative volumes. These are *stress relations at an interface* and *change of stress from point to point*, which are important in themselves and serve as practice with the ideas just summarized.*

STRESSES AT AN INTERFACE

Rocks are full of interfaces: crystal boundaries, bedding planes, fault planes, and plate boundaries are examples. The purpose of this section is to inquire: If we have complete knowledge of the stress state on one side of an interface, how much can we deduce about the stress state on the other side? Part of the answer depends on the nature of the materials and hence is deferred to Chapter 5, but part of the answer can be constructed by assembling ideas we have already used earlier in this chapter. We make progress by considering orthogonal planes, even where the main compression direction in the neighborhood is oblique to the interface in view.

The first idea we used (perhaps unconsciously) is that the normal stresses on the two sides of an interface are equal. At Question 3.4 we imagined a cube with edges 1/8 in. long supporting an automobile by exerting a compressive stress of about 1000 MPa. Some extra details are shown in Figure 3.25 but the essential idea continues the same: the normal stresses on the underside of the interface are exactly matched by the normal stresses on the top side. The same idea appears in Q 3.9 about a floating ice block: at the

---

*Notation: The convention used in paragraph 4 is an example of the "Einstein summation convention." According to this widely used convention, any time that one index appears twice in the same term, that term stands for the sum of three terms. In the example in paragraph 4, the index j appears twice, once in the denominator and once in the numerator; then by the convention, we create three terms with j taking the values 1, 2, and 3 in succession and add those terms. A very common second example is the statement that "mean stress = $\sigma_{ii}/3$." Here the index i is repeated, and hence the statement stands for the more straightforward form "mean stress = $(\sigma_{11} + \sigma_{22} + \sigma_{33})/3$." Because of the way the convention works, σ_{jj} and σ_{kk} both mean exactly the same as σ_{ii}.

A related piece of terminology is the "Kronecker delta" δ_{ij}. Here the conventional meaning is that $\delta_{ij} = 1$ if $i = j$, but $\delta_{ij} = 0$ if $i \neq j$; in other words, only δ_{11}, δ_{22}, and $\delta_{33} = 1$. When readers turn to more advanced texts, they are likely to find both these conventions in use.

FIGURE 3.25

underside of the block, the normal stress exerted upward by the water on the ice is matched by the normal stress exerted downward by the ice on the water.

The second idea, equally simple, is that the tangential stresses on the two sides of the interface must match. The first example of this was a sliding boulder, Question 3.21; whatever shear stress the boulder exerts on the pavement, an equal shear stress is exerted by the pavement on the boulder. The simplicity and power of this idea needs to be emphasized: it does not matter whether the boulder is moving or still, whether it is rough or smooth, or how much normal stress is pressing the boulder and the pavement to-gether—whatever shear stresses are present, whether large or small, the same shear stress is present on the two sides of the interface. The equality is still exactly maintained even when the two materials that meet are very different. For example, remember Question 3.28, about water flowing through a pipe: the shear stress exerted by the water on the pipe wall is equal to the shear stress exerted by the pipe wall on the water. It is the response to the shear stress that is so different in the two materials, not the shear stress itself.

By contrast, the third element of the stress state does not need to be the same on the two sides (see Figure 3.26). We consider an element so small that any internal effect such as the element's own weight is negligibly small: then, as already noted, it must be true that $N_1 = N_2$ and $S_1 = S_2$; but there is no requirement at all that $P_1 = P_2$. As the element we consider becomes even more small, P_{1A} and P_{1B} have to become even more exactly equal, and similarly for P_{2A} and P_{2B}, but there is no law of mechanics that relates P_1 to P_2.

The point just made appears rather clearly in Question 3.30 about boudinage. Figure 3.17 shows successive planes a, b, c, etc.; in answering the problem, we calculated the normal stress on plane b using our estimate of the normal stress on plane a and so on, but the calculations involved only stresses inside the stiff layer. Plane a or plane b could be continued into the surrounding less stiff material, but the normal stress across a plane's continuation is of no concern in the problem; we cannot calculate it from the information we have, nor do we need to know it to work out what is going on inside the stiff layer.

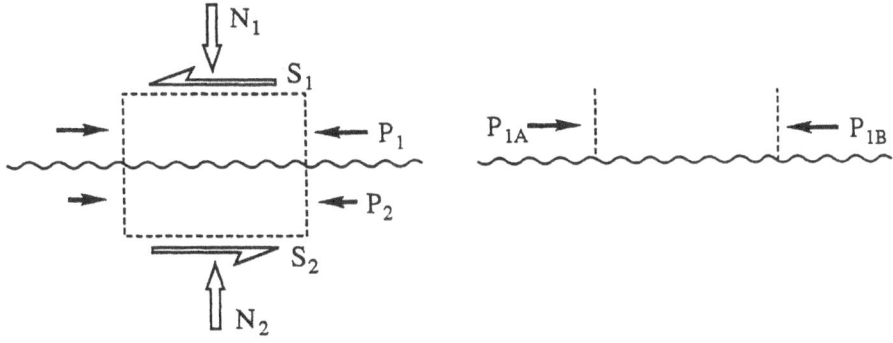

FIGURE 3.26

Question 3.33 Compressive stresses parallel to an interface: Among the problems already done, Question 3.19 concerns an alternating succession of limestone and shale layers and seems to contradict what has just been stated. Given the compressive stress parallel to the layering in the shale, the problem is to calculate the compressive stress in the limestone. What is the extra condition that makes the calculation possible?

Three Dimensions   The discussion of interfaces has been based so far on Figure 3.26, which shows just two dimensions. If one asks what is the three-dimensional situation, including regions in front of or behind the plane of the figure, there are two ideas to note. First, one can always choose to consider a plane normal to the interface for the purpose of making a diagram like Figure 3.26; in fact, if stresses N_1 and N_2 are to be stresses normal to the interface, it is necessary to choose such a plane so as to be able to show them. Second, a complete description can be given by considering *two* planes normal to the interface that are, in addition, normal to each other (Figure 3.27). As noted above, $S_1 = S_2$, but it is not necessary that $P_1 = P_2$; similarly $S_3 = S_4$, but we do not need $P_3 = P_4$. In fact, all four normal stresses P_x are independent of each other, and the pair S_3S_4 is independent of the pair S_1S_2.

If we consider different pairs of orthogonal planes, rotating about the line N_1N_2 as axis, we must necessarily find one pair as in Figure 3.28. It is sometimes important to know the orientation and magnitude of the maximum shear stress, S_{max}: for example, if slip occurs at the interface, the direction along which the material above slips with respect to the material below is usually along the direction of S_{max}, and it is usually the magnitude of S_{max} that determines whether slip occurs or not. On the other hand, the

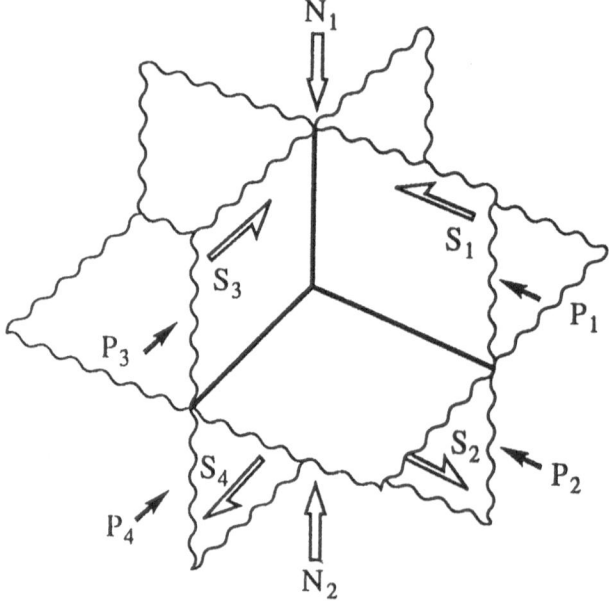

FIGURE 3.27

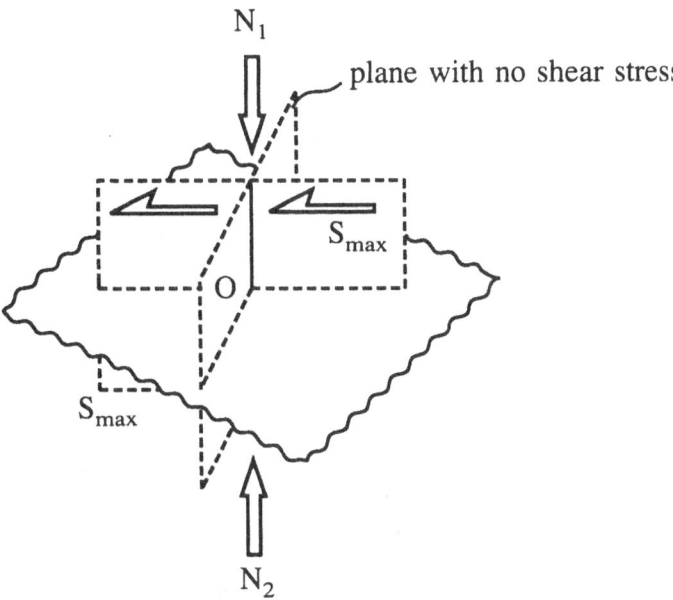

FIGURE 3.28

statements made about the stresses in Figure 3.26 are true for any plane normal to the interface; to use the equations $N_1 = N_2$, $S_1 = S_2$, and $P_1 \neq P_2$, it does not matter whether $S_1 = S_{max}$ or not. (The topic is continued in Chapter 5 as follows: shear stress at an interface, page 114, normal stresses at an interface, page 120, and combination of shear stress and normal stresses, page 128.)

Planes Oblique to Each Other

At the start of the present section, it was emphasized that if two planes are orthogonal, the normal stress on one is independent of the normal stress on the other. The discussion of interfaces just given is an example of this independence: compressive stresses normal to the interface are identical on the two sides, while compressive stresses parallel to the interface are unrelated. But as soon as we consider stresses acting on two planes that are not orthogonal, the independence disappears; in its place, relations emerge that have a charming resemblance to the strain relations. These relations are the topic of the next chapter.

CHANGE OF STRESS FROM POINT TO POINT

The core of the generalizations on page 72 is that although stress in a rock is normally nonuniform, the stress gradients from point to point are related to each other; for example a change in east-west compression along an east-west line might be linked to a change in shear stress in passing from the undersurface of a horizontal bed to its upper surface—the stress cannot change from point to point in arbitrary ways. In parallel with the section on change of strain from point to point, we now look at some simple examples of how stress might vary in real geological situations.

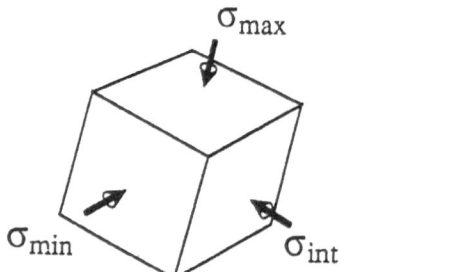 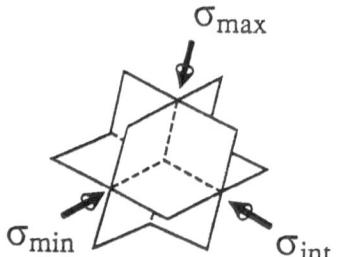

FIGURE 3.29

Stress Fields

At a *point* inside a continuous material, we have a **stress state.** We specify it by giving σ_{max} and σ_{min} and, in three dimensions, by giving $\sigma_{intermediate}$ as well ($\sigma_{intermediate}$ is the normal stress on a plane orthogonal to the planes that carry σ_{max} and σ_{min}; see Figure 3.29). But if we consider a volume of material that is sufficiently large, we notice that the stress state at one end is different from the stress state at the other end. In this situation, we are considering the **stress field,** and in fact noting that the stress field is not homogeneous.

> In a homogeneous stress field, the stress state is the same at every point within the region considered.
>
> In an inhomogeneous stress field, the stress state is *not* the same at every point.

Compare with homogeneous and inhomogeneous strain fields, as in Figure 2.8. Examples of some simple stress fields follow.

The Stress Field in a Fluid at Rest in a Gravity Field   See Figure 3.30: in this figure, at any point, a vertical radius is drawn whose length shows the magnitude of the vertical

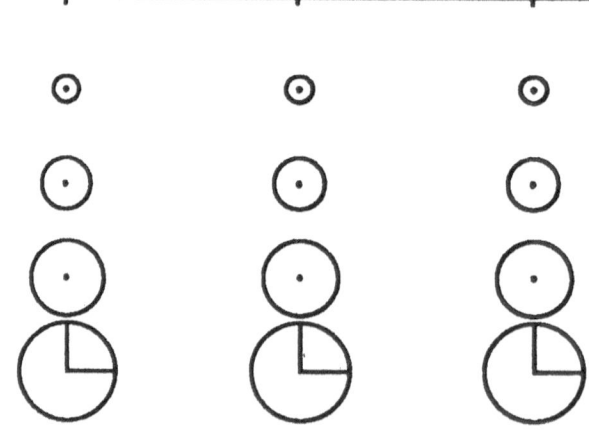

FIGURE 3.30

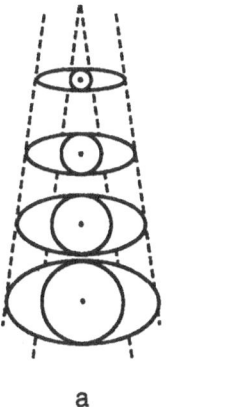 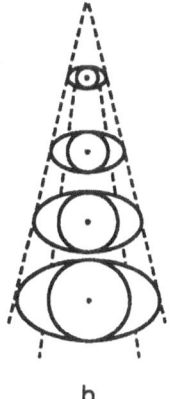

a   b

FIGURE 3.31

compressive stress, and a horizontal radius is drawn whose length shows the magnitude of the horizontal compressive stress. If the fluid is at rest, these two magnitudes are equal (the "hydrostatic" condition) and this equality is emphasized in the figure by drawing circles through the tips of the radii.

The Stress Field in a Fluid Subject to Gravity Plus Extra Horizontal Compression   Two more simple and interesting fields are shown in Figure 3.31. In the stress field in Figure 3.31(a), every horizontal stress is greater than the vertical stress at the same point by a constant *amount,* while in the stress field in Figure 3.31(b), every horizontal stress is greater than the vertical stress by a constant *ratio.* It is not likely that any stress field in the real earth will be quite as systematic as these, but these are intended as reminders of ways in which real stress fields may differ from being hydrostatic or lithostatic.

The Stress Field Beneath an Undulating Topography   This field is linked with a strain field already considered in Chapter 2, Figures 2.37 through 2.40. Beneath point *b* in Figure 3.32, the stress field is of the same general type as the examples in Figure 3.31 with σ_v increasing downward and σ_h larger than σ_v, to cause horizontal shortening. Beneath point *a,* again σ_v increases downward, but here σ_h is *less* than σ_v, allowing horizontal elongation. Around point *f,* σ_v and σ_h are roughly equal, but horizontal and vertical shear stresses are present; the stress state is as shown in Figures 3.32(b) or (c), which are alternative representations of the same state. The stress state has to be of this type to produce the strains shown in Figure 2.40(d); more fundamentally, and regardless

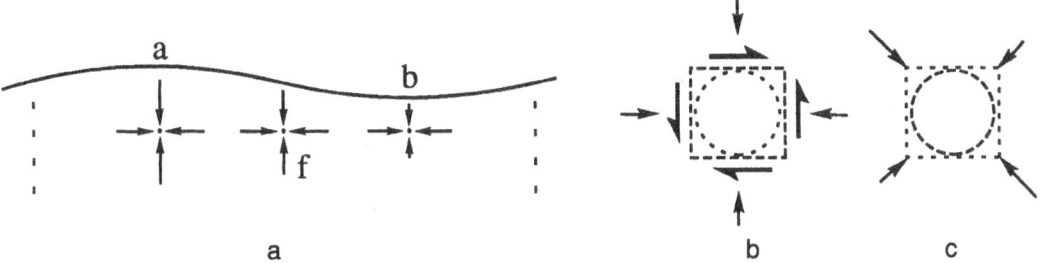

a   b   c

FIGURE 3.32

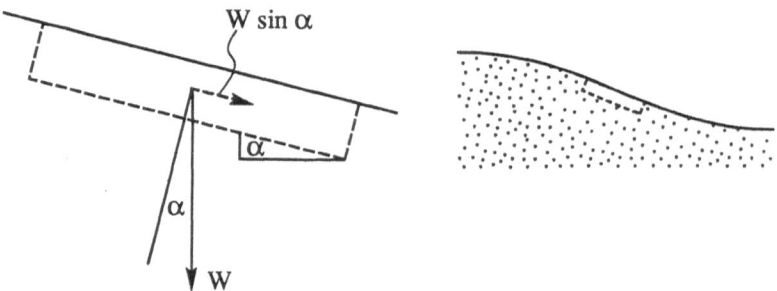

FIGURE 3.33

of any strains, it has to be of this type for the force-balance reasons summarized on pages 71ff.

Sloping hillsides have already been considered at Q 3.12, Q 3.22, and Q 3.23, and will need to be considered again if we try to understand mountain building, so that it is interesting to look at the region between point f and the surface in more detail. A portion of rock at shallow depth below a sloping surface is shown enlarged in Figure 3.33. Its weight has a component parallel to the surface, a force of magnitude $W \sin \alpha$; at Question 3.12 we considered such a slab being kept stationary by unequal pressures normal to its ends, whereas at Question 3.23 we considered such a slab being kept stationary by a shear stress at its base. Beneath a real hillside, such a portion is commonly kept stationary by a combination of shear stress plus not-quite-equal compressions on its ends; but let us ignore the latter effect for a moment and assume that $W \sin \alpha$ is compensated wholly by basal shear stress. By doing this, we find a stress field sufficiently simple to be worth keeping in mind.

The point of interest is that it is possible for all stresses to be proportional to depth below the surface. It is convenient to picture a slab of some specified area A m^2 as in Figure 3.34. Then at density D kg/m^3, the slab's weight is AhD kg. The force component balanced by the basal shear stress is $AhD \sin \alpha$ kg, so that the shear stress itself is $hD \sin \alpha$ kg/m^2. Similarly the force balanced by the basal normal stress is $AhD \cos \alpha$ kg, giving a normal stress on the base of hD cos α kg/m^2. If, in the manner of Figure 3.31(b), the compression acting parallel to the surface also increases linearly downward, the stress field has the character shown in Figure 3.34(c): the proportional increase of all

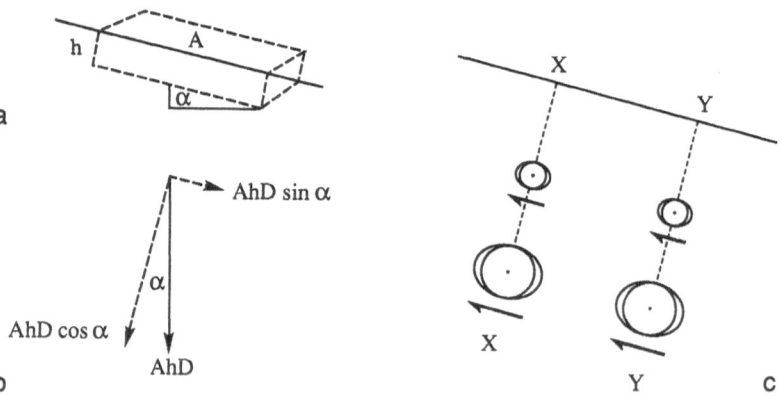

FIGURE 3.34

stresses down line Y is just the same as the proportional increase down line X, or down any other line normal to the surface. As noted, in real life the stress states down line Y are likely to be slightly different from those down line X; also compressions parallel to the surface do not have any obvious reason for increasing downward in such a simple way—they might as easily show some other kind of variation. However, reasons will appear in Chapter 6 that make this proportionality an interesting possibility, and Figure 3.34(c) will be used again in that discussion.

Question 3.34 Stress state beneath a sloping surface: Suppose rock of density 3 g/cm³ or 3000 kg/m³ forms a topographic slope of 6.° For a point beneath such a slope that is 8 km from the surface, and for directions parallel and perpendicular to the slope, as in Figure 3.34(c), estimate the two normal stresses and the shear stress. Assume the compression acting parallel to the slope is 30 percent larger than the compression acting normal to it.

(The topic is continued at Questions 4.9 and 6.12.)

It is important to note that the two normal stresses shown in Figure 3.34(c) and calculated in Question 3.34 are not the maximum and minimum normal stresses for that point; in Figure 3.31(b), the stresses shown *are* the maximum and minimum, but in Figure 3.34(c) they are not. So, pursuing Question 3.34, we might ask: In what direction does the maximum compression act, and what is its magnitude? Such questions are readily answered by methods introduced in Chapter 4.

Summary If a slope is long enough and uniform enough that the stress state at a subsurface point depends only on its distance below the surface (and not on its distance up or down the slope), then for a subsurface plane that is parallel to the surface the following relations hold.

---

normal stress $= hD \cos \alpha = dD \cos^2 \alpha,$ (3.6)

  shear stress $= hD \sin \alpha = dD \cos \alpha \sin \alpha,$ (3.7)

where

$h =$ least distance from the point to the surface;
$d =$ vertical distance from the point to the surface;
$\alpha =$ angle of the surface slope;
$D =$ density of rock.

ANSWERS

Answer 3.1

Two or 3 N. If the apple that fell on Isaac Newton's head had rested there, it would have exerted 1 newton on Newton.

Answer 3.2.

(a) $1/2$ in.²   (b) 3 in.²

(The tire pressure hardly changes when the rider mounts—perhaps by as much as 1 percent if the tire's volume changes by 1 percent.)

Answer 3.3

Roughly the same: a person of 100 kg (220 lb) resting on 100 cm² (the palms of two hands) exerts 1 bar.

Note: The pressure of the atmosphere *is considerable* (easily underestimated). A pressure of 10 bars is hard to imagine.

Answer 3.4

About 10 mm² or a square of side 1/8 in. (Imagine a cube of side 1/8 in. between the jackhead and the auto frame.) It is hard to imagine such a pressure acting on a broader area such as 1 m²; geological pressures are outside our experience.

Answer 3.5

(a)

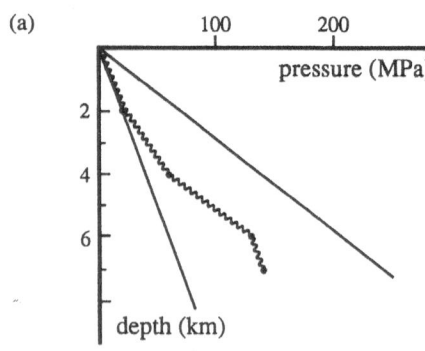

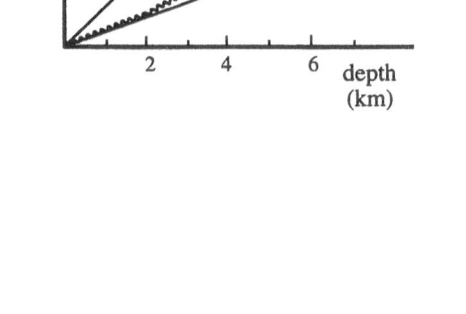

FIGURE 3.35

(b) Where the density is lower the slope is lower. Specifically, with the scales used, slope = 0.3 × density.

(c) 10 MPa   20 MPa   35 MPa   10 MPa

Answer 3.6

In MPa, $dp = 10 \,(2.4)\,(0.0125 \text{ km}) = 0.3$ MPa. (*Check:* does this agree with the orders of magnitude in the table of units?)

Note: The answer just reached is valid regardless of the actual overburden pressure. Whether the layer is close to the surface or 10 km down, the *difference* in pressure from top to bottom must be 0.3 MPa. This must be true because of the rock's density, as long as gravity is as strong as usual, as long as no rock in the vicinity is moving up or down significantly (for the latter effect, see pages 67ff.)

Answer 3.7

Overburden pressure in MPa = $10 \times (2.4) \times$ (depth in km) e.g., at 2 km depth, overburden pressure = 48 MPa; hence formation water pressure = 28.8 MPa. To keep back the formation water, we need at least 28.8 MPa of pressure in the 2 km of drilling mud in the hole, so the needed density is $D?$ where

$$\underset{\text{MPa}}{28.8} = 10 \, (D?) \, \underset{\text{km}}{(2)} \text{ and } D? = 1.44 \text{ g/cm}^3.$$

Answer 3.8

From midpoint to top end = 50 m = 1/20 km = Δh.
Hence Δp in water = 0.5 MPa
and Δp in shale = 1.35 MPa,

or $p_{\text{water}}^{\text{at top}} = p^{\text{midpoint}} - 0.5$
while $p_{\text{shale}}^{\text{at top}} = p^{\text{midpoint}} - 1.35$.
Hence $p_{\text{water}} > p_{\text{shale}}$ by 0.85 MPa or 8.5 bar.

Conclusion: There is considerable potential for the water to split the rock, especially since the water pressure acts on a considerable area (the face of the crack), while the tension in the rock is concentrated much more closely around the tip (Figure 3.36). Salt domes, thunder clouds, and bubbles are blunt at the top as they rise, but a vertical fissure is more like a spear penetrating upward (driven, one may say, by the great pressure exerted by the shale tending to pinch the crack closed in its lower parts).

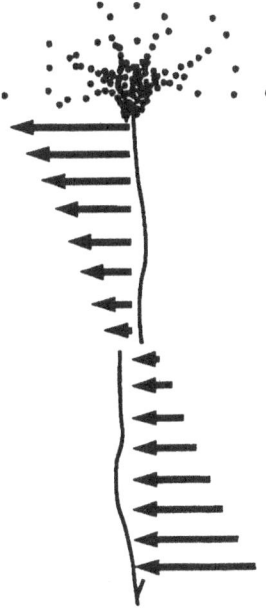

FIGURE 3.36

Answer 3.9

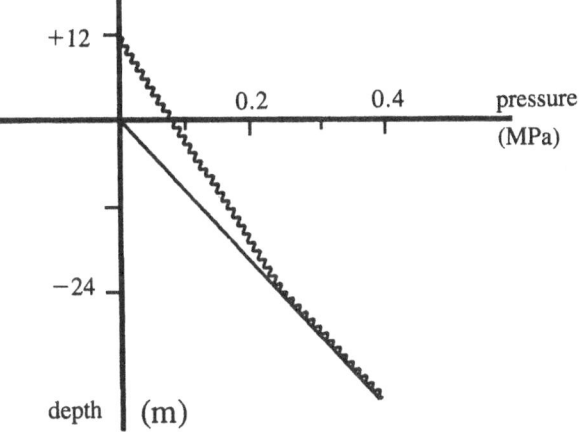

FIGURE 3.37

Answer 3.10

$20 \times 20 \times 70 \times 1.1 = 30,800$ g of force, $= 302$ N on top;

$20 \times 20 \times 90 \times 1.1$ $= 388$ N on base;

difference $= 86$ N, weight of 8800 g, 8.8 kg, around 20 lb (the same as the weight of $(20$ cm$)^3$ of seawater).

Answer 3.11

(a) 3.8 kg or 37 N downward (cube is buoyant).

(b) 1.2 kg or 12 N upward.

(c) 3.8 kg or 37 N in string to anchor; 1.2 kg or 12 N in string to bridge.

Answer 3.12

Force parallel to slope, upward
$= (12$ cm$^2) \times$ (water pressure at left)
$= (12$ cm$^2) \times (15.4$ g/cm$^2)$
$= 184.8$ g.

Force parallel to slope, downward
$= 140$ g $+ (12$ cm$^2) \times (3.3$ g/cm$^2)$
$= 179.6$ g.

Conclusion: The block would roll upward (until water depth at left is only 10.6 cm more than depth on right; then the forces would balance).

Answer 3.13

Upward *force* at front $= 25$ cm$^2 \times 2$ kg/cm$^2 = 50$ kg of force. Hence, upward force at back $= 52$ kg. (Here it is assumed that the emperor's vertical acceleration is zero, and that no other vertical forces are acting.) Hence, average upward pressure at back $= 4$ kg/cm^2.

Note: (i) the core of the reasoning used is TFU = TFD; (ii) the information sought, and some of the data, are *pressures* rather than forces so that the solution process has three legs

```
Pressures   |   Forces
            |
input   - -+-→
            |        ⇓
output   ←-+- -
            |
```

At the dash arrows, we use areas as factors, to convert pressure to force or force to pressure. It is at the open arrow, where we are operating only with *forces,* that we create new information, by using the balance of *forces* TFU = TFD.

Answer 3.14

2 t/m^2 by the method of Answer 3.13.

Answer 3.15

Downward force of salt cube's weight $= 8$ km$^3 \times 2.3$ t/m$^3 = 18.4 \times 10^9$ t.

Shale pressure at 2 km depth $= 2000$ m $\times 2.6$ t/m$^3 = 5200$ t/m^2.

Upward force due to shale pressure $= 5200$ t/m$^2 \times 3$ km$^2 = 15.6 \times 10^9$ t.

TFU = TFD; therefore, force in stem is *upward,* $= 2.8 \times 10^9$ t. If we ask what average pressure exerts this much force, when operating over 1 km^2 of area, average pressure $= 2800$ t/m^2, 280 bars or 28 MPa.

Answer 3.16

Downward force of salt cube's weight $= 18.4 \times 10^9$ t as before.

Upward force due to shale pressure $= 5200$ t/m$^2 \times 3$ 3/4 km$^2 = 15.6 \times 10^9$ t.

TFU = TFD; therefore, force exerted by stem is downward, = 1.1×10^9 t.

Average tensile stress in stem = 4400 t/m², 440 bars, 44 MPa operating over 1/4 km².

Answer 3.17

Considering a representative area of A m², area of pillar cross-sections total 1/4 A m²; area of seam's roof or floor pressed on by water = 3/4 A m².

Total force down = (overburden pressure) × A; total force up = (water pressure × 3/4 A + (pillar pressure) × 1/4 A.

Pillar pressure = 2300 t/m².

Answer 3.18

Consider an area of A m²:

(a) (12 MPa × 0.8 A m²) + (? MPa × 0.2 A m²) = 24 MPa × A m²; therefore, pressure in rock portion = 72 MPa.

(b) 22 MPa
(c) 28 MPa
(d) a state of *tension;* average tensile stress = 8 MPa.

Answer 3.19

(a) 60 MPa × 12/20 A m² + ? MPa × 8/20 A m² = 54 MPa × A m²; thus ? = 45 MPa.
(b) 95 × 12/20 + ? × 8/20 = 54; thus ? = − 7 1/2 MPa—i.e., a tensile stress.

Note: At (b), the limestone beds might part, on account

of the large compressive stress in the shale. Parting in such a situation is called *boudinage,* each limestone segment being one *boudin.* Possible causes for the formation of boudins are discussed further in Questions 3.30, 5.13 and 6.14.

Answer 3.20

Downward force of salt cube's weight = 18.4×10^9 t as before.

Shale pressure at 3 km depth = 7800 t/m²; hence downward force due to shale pressure on cube's top

surface = 4 km² × 7800 t/m² = 31.2×10^9 t.
Total downward force = 49.6×10^9 t.
Shale pressure at 5 km depth = 13,000 t/m²; hence

$$\text{upward force due to shale pressure} = \underset{\text{t/m}^2}{13{,}000} \times \underset{\text{km}^2}{3\ 3/4} = \underset{\text{t}}{48.75 \times 10^9}$$

TFU = TFD; hence force exerted by stem is upward, 0.85×10^9 t.
Average pressure in stem = 3400 t/m², 340 bars, 34

MPa (operating over 1/4 km²). This is only about one quarter of the vertical compressive stress one would expect at 5 km depth.

Answer 3.21

1200 t/m²,   120 bars,   12 MPa

Note: The illustration suggested assuming that the force exerted by ice on the boulder was parallel to the

surface where stress was to be estimated. If this parallelism is not present at the outset, one must generate it by working out a force component that is parallel to the surface of interest, as in Question 3.22.

Answer 3.22

Force parallel to the surface of interest = 1320·cos 65° t.
Stress parallel to the surface of interest = 1200·cos 65° t/m².

Answer 3.23

(a) For some representative area A m², the area of the *rock part* of the porous layer is $0.8\,A$ m². Consider a slab of area A m² lying between planes S and S', and balance forces parallel to the dip direction:

$$\text{force down-dip} = (4 \text{ MPa}) \times (A \text{ m}^2) = \text{force up-dip} = (? \text{ MPa}) \times (0.8A \text{ m}^2)$$
$$\text{on plane } S \qquad\qquad \text{on plane } S'$$

thus, $? = 5$ MPa.

(b) $(30 \text{ MPa}) \times (A \text{ m}^2) = (P_{\text{rock}}) \times (0.8\,A \text{ m}^2) + (P_{\text{fluid}}) \times (0.2\,A \text{ m}^2)$

If $P_{\text{fluid}} = $ negligible, $P_{\text{rock}} = 37.5$ MPa and, reading the graph, the tangential stress the rock can withstand is 5.3 MPa, which is *more* than the stress present.

If $P_{\text{fluid}} = 20$ MPa, $P_{\text{rock}} = 32.5$ MPa and, reading the graph, the tangential stress the rock can withstand is 4.7 MPa, which is *less* than the stress present. In this case, the porous rock is in process of giving way and any people lower on the hill slope should hurry to escape. (Better, one should try to drain such critical units, and prevent the fluid pressure from rising to a magnitude as high as 20 MPa.)

Answer 3.24

TFU = TFD.
Total force down = 2560 g or 25.6 N;

Total force up = 128 cm² × average shear stress. Therefore, average shear stress = 20 g/cm² or 0.2 N/cm².

Answer 3.25

(a) 15 g/cm², or 0.15 N/cm²

(b) 10 g/cm², or 0.10 N/cm²

Answer 3.26

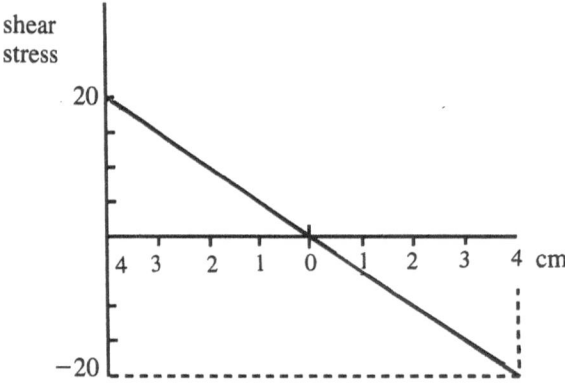

FIGURE 3.38

Answer 3.27

TFU = TFD.
Total force upward = 128 cm² × shear stress on left-hand face, which is 15 g/cm² as at A 3.25(a).
Total force downward = weight 640 g + downward shear force on right-hand face, 128 cm² × 10 g/cm² as at Answer 3.25(b).

Note: Plane P figures in this exercise and in Question 3.25(b), but there we calculated the upward shear force exerted by shaded block on central block, whereas here the force in the calculation is the reverse, the force exerted by central block on shaded block—the same magnitude but now downward.

Answer 3.28

For the plug of water in the pipe,

force pushing to right on end of plug at J = (0.06 MPa) × (cross-section area);
force pushing to left at open end = 0;

resisting force dragging on sides of plug = (? MPa) × (area of the wetted inner surface of pipe $2\pi rL$).

Answer: ? = 48 Pa (0.000048 MPa)

Answer 3.29

6 MPa. Possibly a calculation of this type applies to the lithosphere slab between the Andes and the Mid-Atlantic Ridge (southern part), except that the north and south margins of such a slab might have stresses on them that would also contribute to the balance of forces.

Answer 3.30

For the first 10 cm segment,

force toward left on r.h. end = (23 MPa) × (8 W cm²)   = 184 W MPa-cm²;
force toward right = (6.1 MPa) × (10 W cm²) × 2   = 122 W MPa-cm²,

where the force is due to shear stresses on top surface and underside.

Hence (if acceleration = 0) force toward right from segment (2) = 62 W MPa-cm²

and compressive stress at surface (a) = $7\frac{3}{4}$ MPa. Compressive stress exerted by segment (1) on segment (2) = same, so successive answers for surfaces *b, c, d . . .* , etc, are

b   −3 1/4 MPa—a *tensile* stress
c   −11
d   −16 1/4

e   −19 1/2
f   −21 1/4
g   −22

h   −22 1/4
i   same as g

j   same as f
etc.

These magnitudes are plotted in Figure 3.39. This exercise illustrates the fact that, once boudins form, there is a tendency for each one to split in the middle, especially those that happened to be longer among the first-formed set. (The topic continues in Q 5.3.)

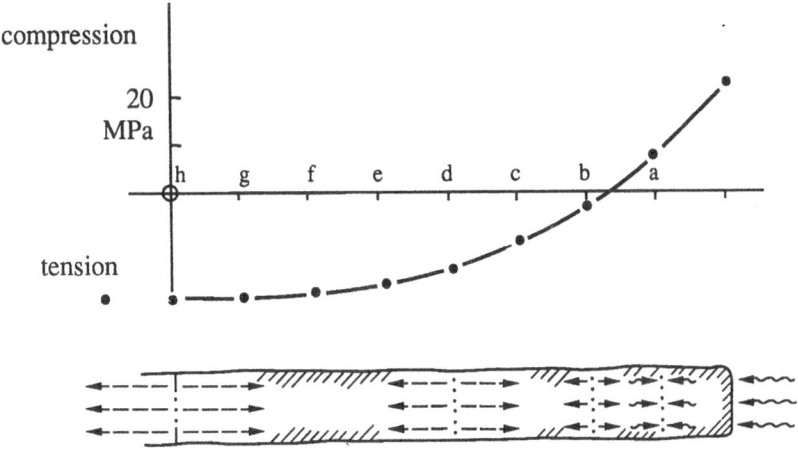

FIGURE 3.39

Answer 3.31

(a) buoyant cube, of weight 5 kg: force by swimmer = 3.8 kg downward, giving an average tangential stress of 0.38 kg/cm², 0.37 bar, 0.037 MPa, downward by swimmer on cube and upward by cube on swimmer.

(b) sinking cube, of weight 10 kg: average shear stress 0.12 kg/cm², upward and downward, or 0.012 MPa.

Answer 3.32

(a) The rectangular root in Figure 3.22(d) can support $\frac{1}{9} \times 5\frac{1}{3}$ km of overburden isostatically, = 0.59 km. The total overburden is 1.33 km so the unsupported part is 0.74 km thick. The unsupported volume is thus $0.74 \times 170 \times L$ km³ or $126\,L$ km³ or $126.10^9\,L$ m³. At 2800 kg/m³, we need an upward force of $2800.126.10^9\,L$ kg.

(b) For the average shear stress, divide the force at part (a) of this question by the area of wall, 80 L km² or $80\,L \cdot 10^6$ m²: the result is 441.10^4 kg/m² or 441 kg/cm² or 433 bar or 43.3 MPa.

(c) The pressure under 40 km of rock of relative density 2.8 is $10 \cdot (2.8) \cdot 40$ MPa or 1120 MPa, so the average shear stress calculated is 4 percent or $\frac{1}{25}$ of this compressive stress.

Answer 3.33

The extra condition is that the average compression for the succession as a whole has to be equal to some stated amount. Given this condition, if the stress in the shale increases, the stress in the limestone must decrease: the two are no longer independent. But the dependence comes directly from the extra condition, together with knowledge of the layer thicknesses. This special case where the two *are* related serves to emphasize that in common geological circumstances they are *not*.

Answer 3.34

Compared with Figure 3.34(b), $D = 3000$ kg/m³, $h = 8000$ m, and $\alpha = 6°$, hence $\cos \alpha = 0.99$ and $\sin \alpha = 0.10$. Basal shear stress = $(2.4)\,10^6$ kg/m² and basal normal stress = 24.10^6 kg/m²; normal stress acting parallel to the slope = 31.10^6 kg/m². In more commonly used units, these are 24, 240, and 310 MPa or 0.24, 2.4, and 3.1 kb respectively.

Note: Two significant figures are enough for expressing geological conclusions, but to maintain the factor of 30 percent, the more exact values 24, 238, and 309 MPa can be used.

Variation of Stress with Direction

Mountains and outcrops are more agreeable to look at than city streets. Part of the virtue is the variety of texture—the spongy moss, the craggy cliff, the delicacy of the thin little stream trickling down through the rocks. Another part is the variety of attitude: in a city street, most surfaces are vertical or horizontal, rather monotonous, whereas rocks show an abundance of planes, and every attitude one can imagine—a sweet disorder. But of course the planes are not totally random; there is some underlying system. It is the business of geomorphology, rock mechanics, and structural geology to find reasons and causes for the obliquities we see.

STRESS ON OBLIQUE PLANES

The kernel of Chapter 2 was the idea that strain at a point is specified by three numbers (or two if we limit ourselves to two dimensions), but that the three numbers specify an infinite set of individual related quantities—the linear strain and shear strain components for any direction through the point. The Mohr circle and the sine-wave diagram are both devices for displaying the systematic range of individual components that all coexist with the three principal strains. The purpose of the present section is to find similar relations among stresses on different planes through a single point. As elsewhere, we begin with a numerical example and then go to more abstract and general ideas.

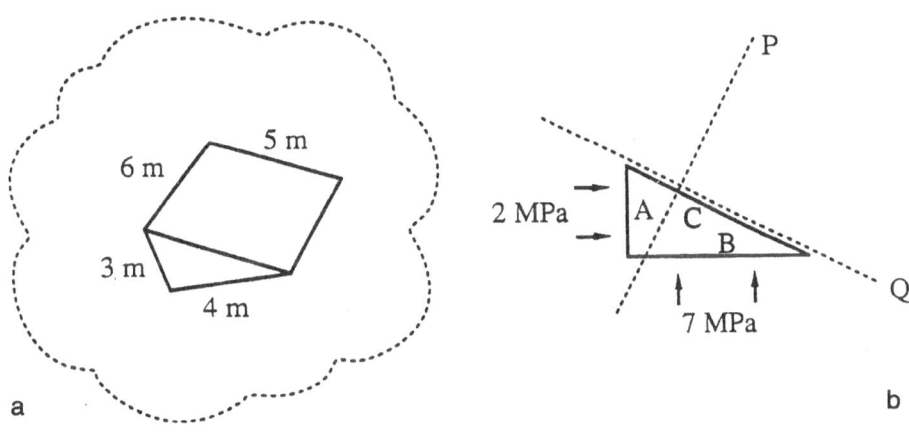

FIGURE 4.1

Question 4.1 Balance of forces on a wedge: Figure 4.1 shows a wedge-shaped element in the interior of a mass of rock, and two faces *A* and *B* that have no shear stress on them—only normal stresses. Our objective is to discover what stresses must necessarily be present on the inclined face *C*. Our procedure is to balance forces (TFU = TFD) first along direction *P* and then along direction *Q*.

(a) What is the area of face *A* (extending 6 m perpendicular to the page)?
What force is meganewtons acts on this face? (One meganewton, MN, is exerted if 1 MPa acts on 1 m².)
What is the *component* of this force along direction *P*? (*Suggestion:* Multiply by the cosine

of some angle, which depends on the dimensions in Figure 4.1 (a).)

(b) What is the area of face *B*?
What force in MN acts on face *B*?
What is the component of this force along direction *P*?

(c) What single force acting down along direction *P* will balance the effects at (a) and (b) combined?

If the wedge is weightless and not accelerating, a force as calculated at (c) must be present somewhere.

(d) Suppose the needed force comes from a normal compressive stress on face *C*: What is the area of face *C*? What is the stress magnitude in MPa? (Answer: 5.2 MPa)

Question 4.2 Forces on a wedge, continued:

(a) For the normal force on face *A*, what is its component along direction *Q?*

(b) For the normal force on face *B*, what is its component along direction *Q?*

(c) What is the combined effect from (a) and (b)

along direction *Q?* (Add or subtract as needed.) What single force acting tangentially to face *C* will balance the combined effect?

(d) Suppose the needed force comes from a shear stress on face *C:* What is the needed shear-stress magnitude? (Answer: 2.4 MPa)

In working the preceding exercises, the first thing to do is to get the right answers. But then there are several points that this specific example illustrates, as follows.

The Calculation Process   Note that we are proceeding again in the manner of Question 3.13 about the chair-borne emperor:

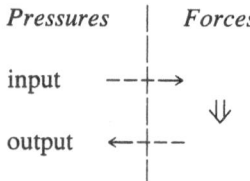

The information we start from is about stresses, and so is the information we seek, but no law of mechanics relates a stress to another stress: the essential law of mechanics relates forces (or conserves momentum). Thus we use area factors at Question 4.1(a) and (b) to get some information about forces, then manipulate the force components and create a balance of forces; the last thing we do is to use an area factor again to convert the information back into stresses.

Stresses at a Point   In Question 4.1 the weight and acceleration of the wedge were ignored and the balance of forces is in error by that amount, but the weight and acceleration becomes less important if we consider a smaller wedge. Consider a hand-sized wedge instead, for example measuring 3 cm by 6 cm by 4 cm: each length is smaller by $100\times$; each area is smaller by $10^4\times$; so each stress-related force is smaller by $10^4\times$; but the volume and weight are smaller by $10^6\times$. Thus the *relative* importance of the weight is less, compared with the stress-related forces. And clearly, by thinking about a wedge measured in micrometers or nanometers, we could make the effect of the weight as insignificant as we wish. In the limit, we reach a conclusion about the stress relations "at a point." If, at a point, there are normal stresses of 2 MPa and 7 MPa as in Figure 4.1, there must also be normal and shear stresses of 5.2 MPa and 2.4 MPa on a third plane through the same point. The result does not depend on assuming that the rock is weightless: the relations among stresses at a point are exactly true no matter how heavy or dense the material may be.

Area Ratios   It is only in the form of ratios that the areas of the wedge faces enter the calculation. If we change the length of the wedge from 6 m to 60 m, the answers would come out the same. For example, at Question 4.1(a) we multiply by 18 m², but at (d) we divide by 30 m² so that in effect we just use the area ratio 3/5, which is the cosine of one of the angles, the same cosine that we used to establish the magnitude of one of the force components. Thus we come to a basic difference between stress relations and force-component relations: in the latter, just a simple cosine relates one component to another, but in stress relations the trigonometrical factors come in twice—once as area ratios as well as once in evaluating force components. To illustrate this point, let us repeat Question 4.1 with a 45° wedge: cosine 45° = $1/\sqrt{2}$, an irrational number, but no $\sqrt{2}$ will appear in the answers, because every term will involve either $(\sqrt{2}) \times (\sqrt{2})$ or $(\sqrt{2})/(\sqrt{2})$—the trigonometry appears twice over.

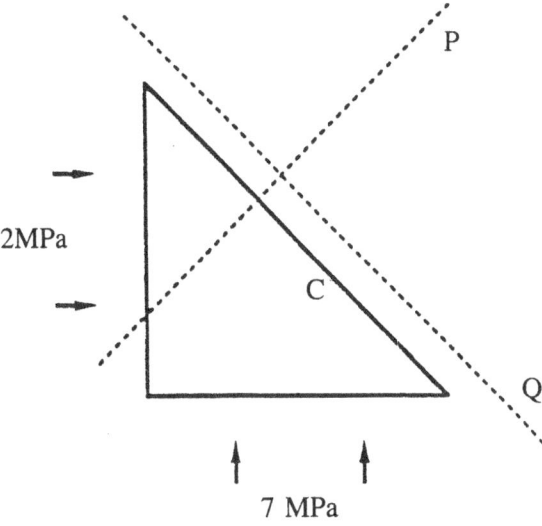

2MPa

7 MPa

FIGURE 4.2

Question 4.3 Forces on a wedge, continued: In Figure 4.2 what stress acts normal to face C, and what shear stress acts tangentially to it? (*Suggestion:* Give face C some arbitrary area, say, S m².)

Next for discussion are two geological examples. Readers can explore these examples using the procedures introduced, which are direct and concrete, and explicitly balance forces measurable in meganewtons. Alternatively, readers can skip to the next section, where abstract general formulas and diagrams are introduced—convenient tools for all questions of the type in hand; and, having grasped the general-purpose tools, can come back to these two specific situations as examples.

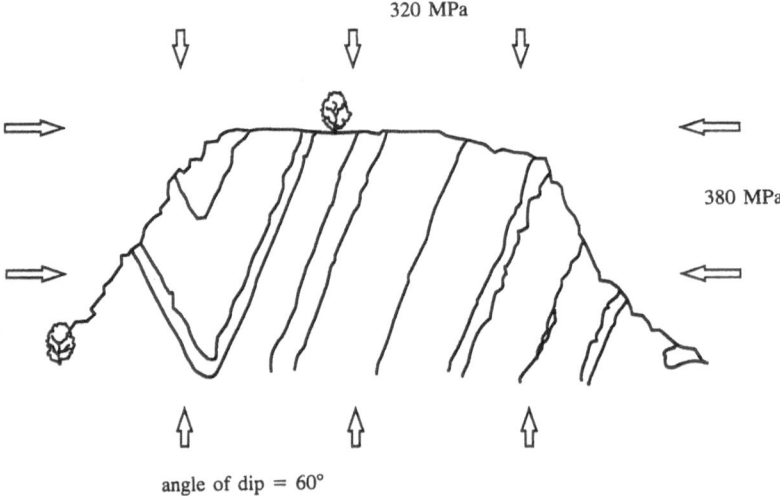

320 MPa

380 MPa

angle of dip = 60°

FIGURE 4.3

Question 4.4 Stresses on an inclined plane in rock: Assume that the rock in the outcrop picture, Figure 4.3 was formerly at considerable depth in the earth, where it was subjected to uniform stresses 320 MPa and 380 MPa as shown.

(a) At roughly what depth would such a vertical-stress magnitude be expected?

(b) What are the normal stress and shear stress on one of the dipping interfaces (probably original bedding)?

(c) Refer to Figure 3.11 but multiply all the stress magnitudes shown by 10. If the curve then shows stress combinations at which the rock in Figure 4.3 fractures, will it fracture under the stress you calculated at (b)?

Question 4.5 Stresses in rock, continued: Sketch in a plane dipping at 30°, perpendicular to the bedding in Figure 4.3. Is fracturing more likely to occur on planes with this attitude?

From the preceding numbers, one might conclude that the rock illustrated would not fracture under these stresses, but to do that would be to ignore the effect of fluids in the material. It is not until the last section of Chapter 6 that we get to the crux of the present problem.

335 MPa average
345 MPa in shale

FIGURE 4.4

Question 4.6 Stresses in rock, continued: In the preceding question, a normal-stress magnitude was calculated for a plane cutting across the layering in Figure 4.3. But the layers are presumably of different compositions; even at the simplest, two lithologies might alternate, as in Question 3.19 about shale and limestone. If the layering were 9/10 shale and 1/10 quartzite as in Figure 4.4, and the layer-parallel compression were 345 MPa in the shale, might the quartzite fracture?
To decide whether the stress magnitude in the quartzite is likely to be so low, we need to know its deformability, as discussed in Chapter 5, especially Question 5.11; then we can go on to consider the likelihood of parting, as in Chapter 6. It is important to remember that the wedge calculations assume a *uniform average* stress over each surface. Any time we use wedge reasoning to relate one uniform average stress to another, it is proper to follow up with a supplementary inquiry: What actual local stresses make up the uniform average we just calculated?

Summary If we consider two planes at right angles in the interior of a continuous sample of material, the normal stress on one is independent of the normal stress on the other; there is no simple general relation between them. But given the two values of the two normal stresses, any third oblique plane's stresses are related to the first two.
    The reason for relations existing is that forces must balance; or in other words, whatever direction we choose as "up," TFU = TFD.
    If we consider stresses on tiny planar elements "at a point," the relations among them are not affected by the material's weight or acceleration, nor by the material's response to the stresses. The material may be swelling, shrinking, flowing, bouncing, or responding in some other way but none of that has any effect on the relations among stresses.
    In the examples considered so far, the two orthogonal planes were assumed to have no shear stresses on them. If shear stresses are present, relations among the stresses continue to exist and to have the same general form but are not so easy to explore by simple arithmetic examples; the diagrams in the next section are a more effective approach.

MOHR CIRCLE AND SINE-WAVE DIAGRAM FOR STRESSES

We can make rapid progress by using precisely the method of Answers 4.1 on the wedge-shaped element in Figure 4.5. Force normal to face $f = \sigma_1 M \cos \theta$, with force components $(\sigma_1 M \cos \theta) \cdot \cos \theta$ along P and $(\sigma_1 M \cos \theta) \cdot \sin \theta$ along Q. Force normal to face $g = \sigma_2 M \sin \theta$, with force components $(\sigma_2 M \sin \theta) \cdot \sin \theta$ along P and $(\sigma_2 M \sin \theta) \cdot \cos \theta$ along Q.

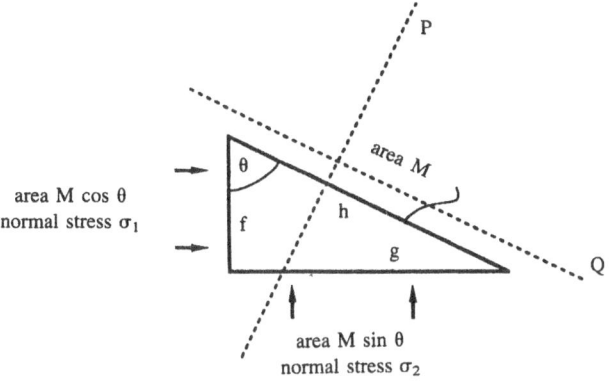

FIGURE 4.5

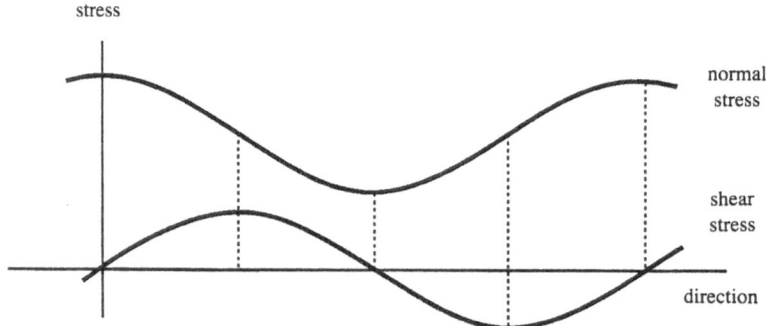

FIGURE 4.6

Along P, total force $= \sigma_1 M \cos^2 \theta + \sigma_2 M \sin^2 \theta$, so stress normal to face $h = $ force/area $= \sigma_1 \cos^2 \theta + \sigma_2 \sin^2 \theta$.

Along Q, total force $= \sigma_1 M \sin \theta \cos \theta - \sigma_2 M \sin \theta \cos \theta$, so shear stress tangential to face $h = (\sigma_1 - \sigma_2) \sin \theta \cos \theta$.

The results are even more compact if we use the double angle 2θ and the relations

$$\sin \theta \cos \theta = \tfrac{1}{2} \sin 2\theta,$$
$$\sin^2\theta = \tfrac{1}{2}(1 - \cos 2\theta),$$
$$\cos^2 \theta = \tfrac{1}{2}(1 + \cos 2\theta).$$

Then on face h,

$$\text{normal stress} = \tfrac{1}{2}(\sigma_1 + \sigma_2) + \tfrac{1}{2}(\sigma_1 - \sigma_2) \cos 2\theta, \tag{4.1}$$

$$\text{shear stress} = \tfrac{1}{2}(\sigma_1 - \sigma_2) \sin 2\theta. \tag{4.2}$$

Before trying to use these questions, let us consider how the normal stress changes if we make θ a smaller angle. As θ approaches zero, 2θ approaches zero as well; $\cos 2\theta$ approaches 1 and the normal stress on plane h approaches σ_1. In fact, σ_1 is the maximum normal stress anywhere in the rock because 1 is the maximum value $\cos 2\theta$ can have.

A similar piece of reasoning shows that σ_2 is the minimum value for the normal stress anywhere in the rock, so that we can rewrite the equations:

$$\text{normal stress} = \tfrac{1}{2}(\sigma_{max} + \sigma_{min}) + \tfrac{1}{2}(\sigma_{max} - \sigma_{min}) \cos 2\theta; \tag{4.3}$$

$$\text{shear stress} = \tfrac{1}{2}(\sigma_{max} - \sigma_{min}) \sin 2\theta. \tag{4.4}$$

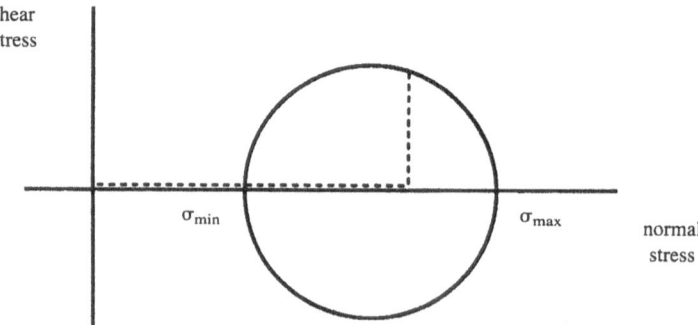

FIGURE 4.7

The correspondence with Equations (2.18) and (2.19) is striking; the only difference is the factor of ½ on the left of (2.19). Necessarily a sine-wave diagram for stresses is also an exact match, as in Figure 4.6; and also a Mohr circle diagram must exist with center at $\frac{1}{2}(\sigma_{max} + \sigma_{min})$ and radius $\frac{1}{2}(\sigma_{max} - \sigma_{min})$, as in Figure 4.7.

Question 4.7 Diagrams for rock stresses: Draw a Mohr circle and a sine-wave diagram to illustrate Questions 4.1 and 4.2, where σ_{max} = 7 MPa and σ_{min} = 2 MPa.

(a) Your answers at Questions 4.1(d) and 4.2(d) should fix a single point on the Mohr circle and two points on the same vertical line in the sine-wave diagram. Check these suggestions.

(b) In Figure 4.1 the oblique plane makes an angle of about 37° with the plane compressed by σ_{max}. Does this angle at 37° appear in its proper place in the sine-wave diagram? Does an angle 2θ (i.e., 74°) appear in the Mohr diagram?

(c) Do your answers to Question 4.3 check?

Question 4.8 Diagrams for rock stresses, continued: Draw a Mohr circle to illustrate Question 4.4 (*Note:* It is not necessary to show normal stresses all the way down to zero; a segment of normal-stress axis running from 300 to 400 MPa is all you need.) Check your answer to Question 4.4(b) and your stress magnitudes from Question 4.5.

The exercise just completed brings up a point that has wide application: the stresses on two planes that are 90° apart in real space will be represented by points that are 180° apart in a Mohr diagram. That is to say, the stress states will be represented by points at the opposite ends of a diameter or, if we start by plotting the two stress points, the *center* of the circle they lie on must be where the line joining them cuts the normal-stress axis. The following exercise is a numerical application of this idea.

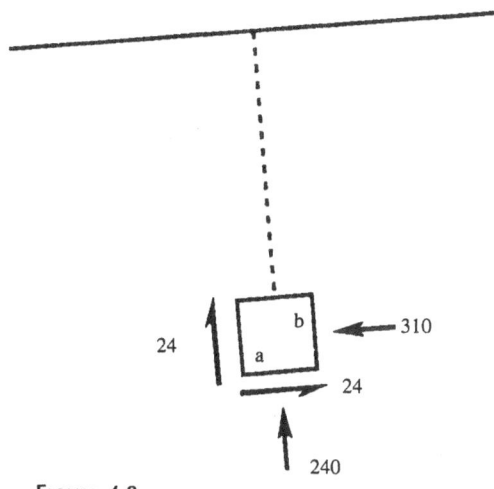

FIGURE 4.8

Question 4.9 Principal stresses: The stress state calculated at Question 3.34 is shown in Figure 4.8. What are the magnitudes and orientations of the principal stresses?

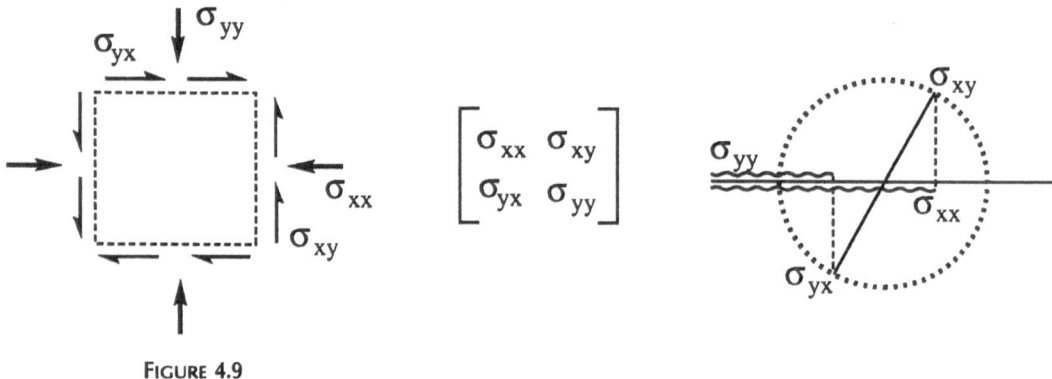

FIGURE 4.9

Summary of Stress Relations

Connection with Mohr Circle   From §5 on p. 73, reduce to two dimensions and use x, y in place of x_1, x_2:

$$\begin{bmatrix} W_x \\ W_y \end{bmatrix} = \begin{bmatrix} \frac{\partial}{\partial x} & \frac{\partial}{\partial y} \end{bmatrix} \begin{bmatrix} \sigma_{xx} & \sigma_{xy} \\ \sigma_{yx} & \sigma_{yy} \end{bmatrix}$$

The three representations in Figure 4.9 are equivalent. Each is a representation of the complete stress state and implies what you would see if you chose any other pair of reference directions for axes.

There is the special case where we choose the principal directions as axes, as shown in Figure 4.10. As with strains, the entire set of possibilities can be represented by an ellipse or by a complete Mohr circle or by a sine-wave diagram.

REFRACTION OF STRESS

A simple but important application of the ideas just introduced deals with stresses at an interface between two rocks. The topic is treated on page 74 but can now be discussed more fully, as in the following examples.

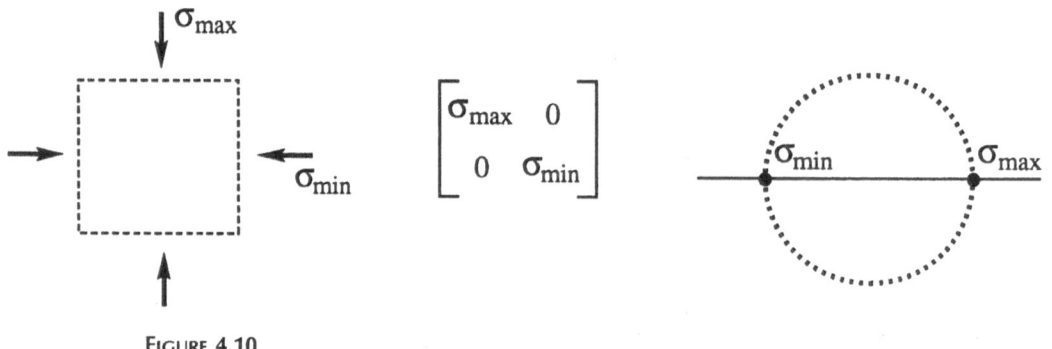

FIGURE 4.10

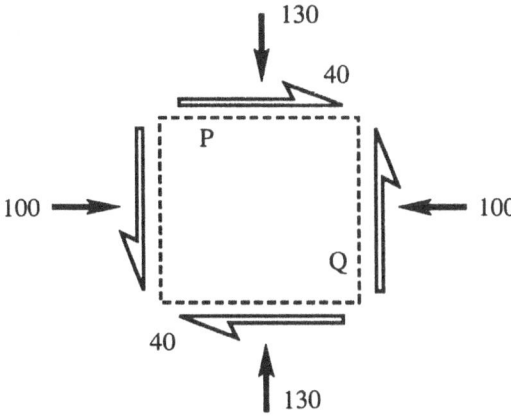

FIGURE **4.11**

Question 4.10 Stresses at an interface:

(a) In the stress state, Figure 4.11, what is the shear stress on plane *Q?*

(b) Plot a Mohr diagram with a point above the horizontal axis to represent the stress on plane *P,* and a point below the axis to represent the stress on plane *Q.* Recall why these points must lie on a *diameter* of the stress circle; hence draw a circle for the stress state shown.

(c) Let θ be the angle between plane *P* and the plane in the rock that suffers σ_{max}. Find 2θ in the diagram and so estimate θ.

(d) Note that to produce the shear stresses shown, the direction of σ_{max} must lie in the NW and SE quadrants (if σ_{max} lay in the NE and SW quadrants, the shear stresses would be of the opposite sense). Using your answer at (c), show the direction of σ_{max} and a plane that suffers σ_{max} in Figure 4.11.

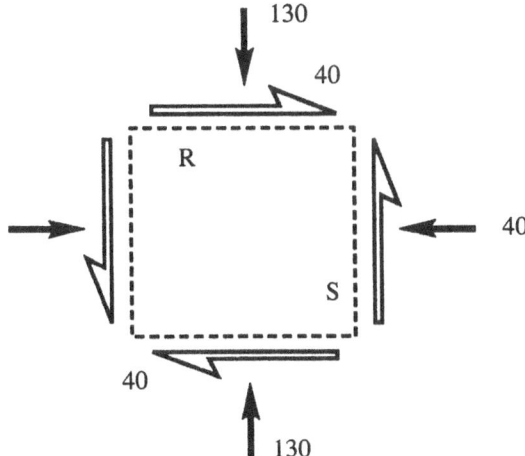

FIGURE **4.12**

Question 4.11 Stresses at an interface, continued: Repeat Question 4.10 for the stresses in Figure 4.12.

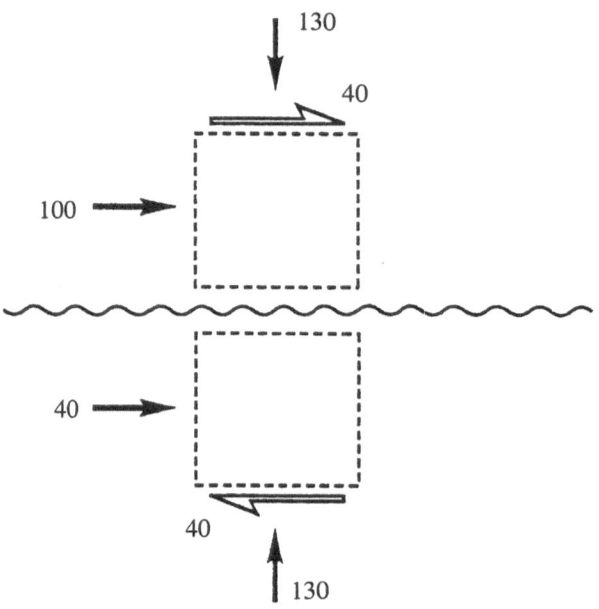

FIGURE 4.13

Question 4.12 Stresses at an interface, complete picture: Suppose the stress states in the two preceding questions are present on the two sides of an interface as shown in Figure 4.13. Add lines to Figure 4.13 to show just the directions of σ_{max} on the two sides of the interface.

Generalizations

1. The stress states on the two sides of an interface are always represented by Mohr circles that intersect; one point of intersection shows the normal and shear stress magnitudes at the interface (e.g., Figure 4.25 in Answer 4.11).

2. If the orientation of the principal stresses is needed, the Mohr-circle description of the stress states is convenient; but if the material response is the thing needed (i.e., linear and shear strain rates), the separate components parallel and normal to the interface are more convenient to work with than principal stresses. As will be seen, one can proceed directly from Figure 4.13 to the material response, page 128, without sidetracking to find the principal directions.

STRESS RELATIONS IN THREE DIMENSIONS

At the end of Chapter 2 it was shown that any set of three strain-rate numbers could be partitioned into a mean value and three remainders that total zero; further, if needed, the remainders could be expressed as two plane strains. In an exactly similar way, any set of three principal stresses can be partitioned into a mean value and three remainders that total zero. For example, a stress state with principal values 330, 270, and 240 MPa can be thought of as combining a mean stress of 280 MPa and remainders of 50, −10, and −40 MPa; if necessary the remainders can be thought of as containing two subsets (10, −10, 0) and (40, 0, −40). The action of each of the parts is easy to imagine, each subset can be shown by a Mohr circle, and the action of the complete system is just the sum of the parts. If the stress acts on a sample of material that deforms homogeneously and does not change volume, the two subsets are convenient for describing the material's response. But if the material is able to change volume—for example, by losing pore fluid or by some part of the sample going into solution and slipping away—partitioning the stress state into subsets is much less useful. These aspects are explored in Chapter 5.

ANSWERS

Answer 4.1

(a) $3 \times 6 = 18$ m²; $18 \times 2 = 36$ MN;
    $36 \times 3/5 = 21.6$ MN
(b) $4 \times 6 = 24$ m²; $24 \times 7 = 168$ MN;
    $168 \times 4/5 = 134.4$ MN

(c) 156 MN
(d) 30 m²; $156/30 = 5.2$ MPa

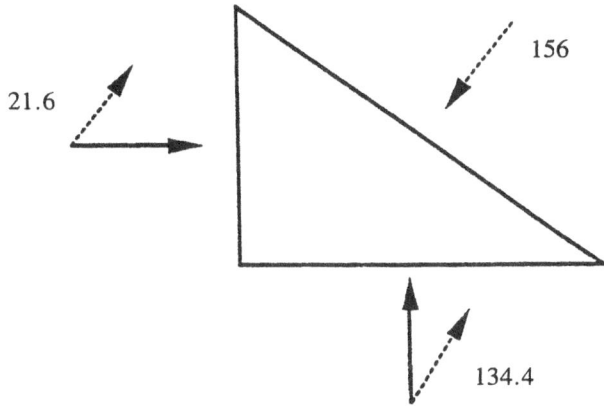

FIGURE 4.14

Answer 4.2

(a) $36 \times 4/5 = 28.8$ MN
      (c) 72 MN

(b) $168 \times 3/5 = 100.8$ MN
      (d) $72/30 = 2.4$ MPa

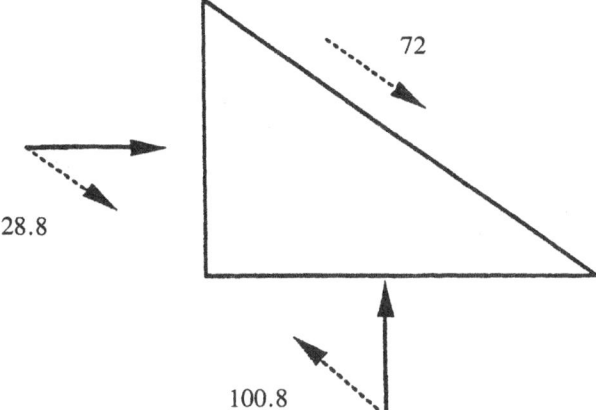

FIGURE 4.15

Answer 4.3

Let face C have area S m²; then the other two faces have areas $S/\sqrt{2}$ m².
Then forces are $2S/\sqrt{2}$ and $7S/\sqrt{2}$ MN.
Force components are $1/\sqrt{2}$ × force acting, = S at side and 3½ S at base.

Along P, net force = 3½ S + S = 4½ S; stress = 4½ MPa.
Along Q, net force = 3½ S − S = 2½ S; stress = 2½ MPa.

Answer 4.4

(a) 10−12 km depth, 1/3 to 1/2 × thickness of crust.
(b) Force A = 380 (0.866S) MN, with components 380 (0.866S) · (0.866) and 380 (0.866S) · (0.5),
    Force B = 320(0.5S) MN, with components 320 (0.5S)(0.866) and 320 (0.5S)(0.5).
    Normal force on interface = 380 S (0.866)² + 320 S(0.5)² = 365 S MN;

normal stress on interface = 365 MPa.
Tangential or shear force on interface = 380 S (0.866)(0.5) − 320 S (0.866)(0.5) = 26 S MN;
tangential or shear stress on interface = 26 MPa.
(c) Rock will *not* fracture under the combination of stresses calculated at (b).

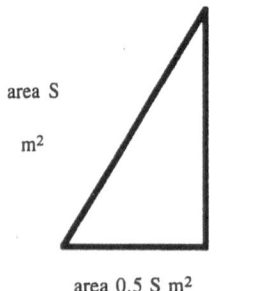

area S m²

area 0.866 S m²

area 0.5 S m²

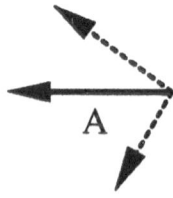

A

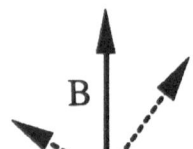

B

FIGURE 4.16

Answer 4.5

Force C = 380 (0.5S) with components 380 (0.5S) (0.5) and 380 (0.5S)(0.866).
Force D = 320 (0.866S) with components 320 (0.866S) (0.5) and 320 (0.866S)(0.866).
Normal force on interface = 380 S (0.5)² + 320 S (0.866)² = 335 S MN;
    normal stress on interface = 335 MPa.
Tangential or shear force on interface = 380 S

(0.5)(0.866) − 320 S (0.5)(0.866) = 26 S MN, the same as before;
tangential or shear stress on interface = 26 MPa.
Fracturing is slightly more likely (assuming the curve in Figure 3.11 still applies), but is still nowhere near being expected (i.e., fracture is slightly less unlikely).

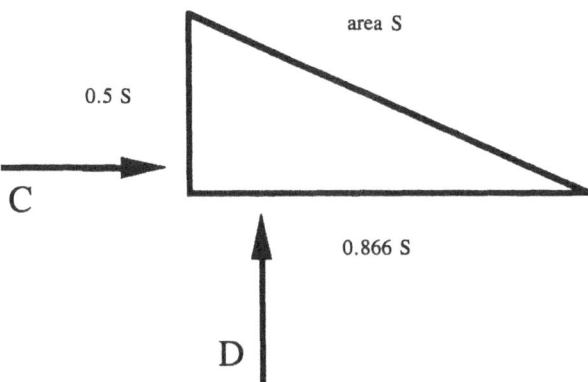

FIGURE 4.17

Answer 4.6

Let the compressive stress in the quartzite layers be x MPa. Then $(1/10) \cdot x = (9/10) \cdot 345 = 335$ MPa, so $x = 245$ MPa. The *shear* stress in the quartzite as shown in Figure 4.18 is the same as in the shale, $= 26$ MPa. If we extrapolate the curve in Figure 3.11 to make Figure 4.19 we find that the quartzite could be very close to fracturing.

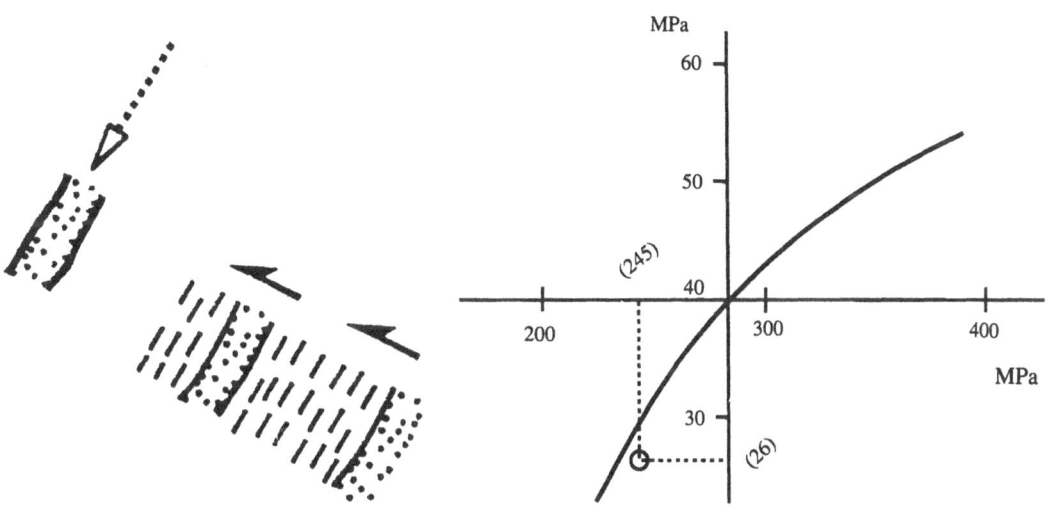

FIGURE 4.18   FIGURE 4.19

Answer 4.7

See Figure 4.20.

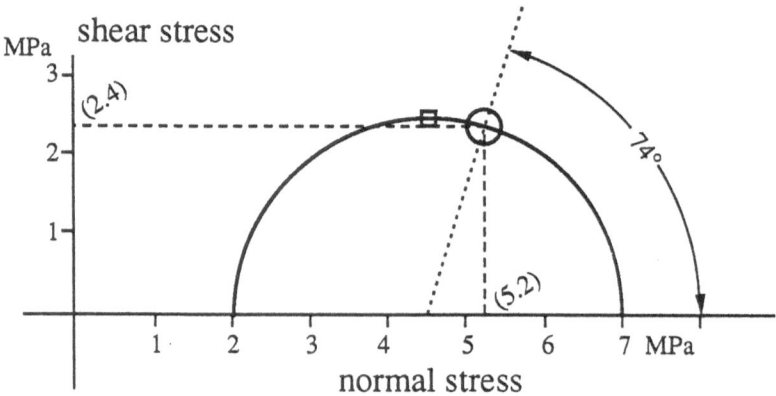

a

FIGURE 4.20A

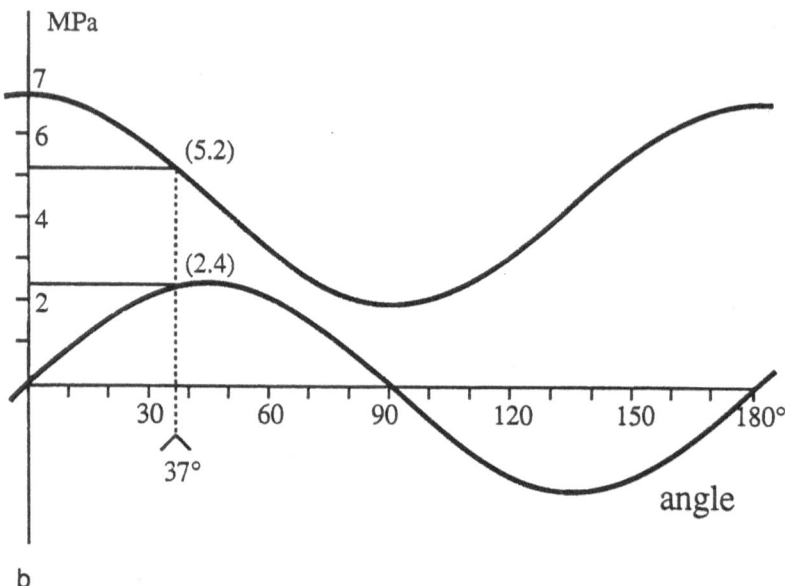

b

FIGURE 4.20B

Answer 4.8

See Figure 4.21. Why use the bottom half of the circle? The two planes we are looking at make 90° in the outcrop, so to have them at 180° to each other in the Mohr diagram keeps everything on track.

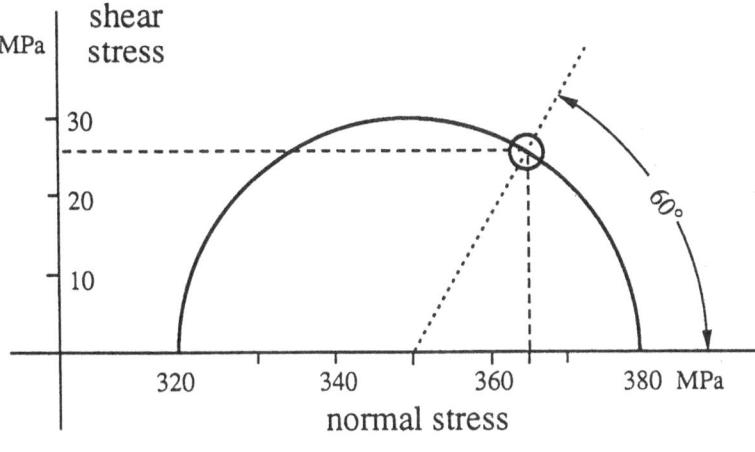

a

FIGURE 4.21A

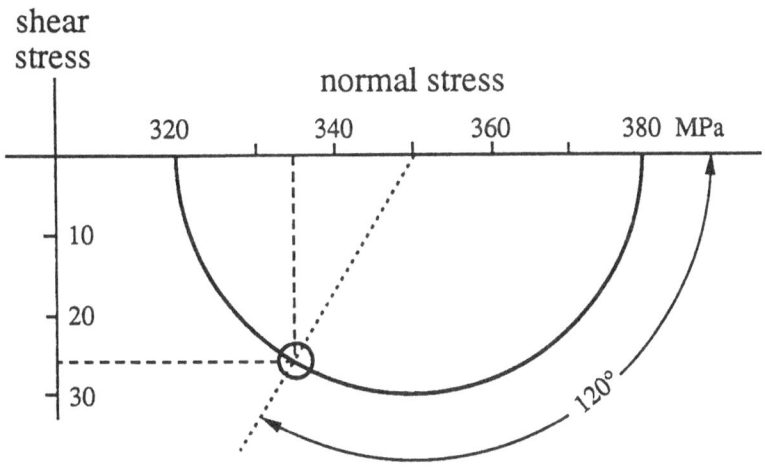

b

FIGURE 4.21B

Answer 4.9

See Figure 4.22. As in Question 4.8, we plot the normal and shear stresses on face (a) from Figure 4.8 at point A and the normal and shear stresses on face (b) at point B. Line AB is a diameter of the needed stress circle; $\sigma_{max} = 317$ MPa, $\sigma_{min} = 233$ MPa. The plane on which σ_{max} acts makes $1/2 \times 34°$ or $17°$ with plane (b) and $23°$ with the vertical.

Note: The Mohr circle diagram shows clearly a 2θ of $34°$, corresponding to a real divergence of $17°$, but how does one decide whether to measure the $17°$ clockwise or counterclockwise from face (b)? One way is to set up a sign convention and adhere to it, but a more robust, physics-based method is as follows. Consider the shear stresses, of magnitude 24 MPa; to create these by themselves, we would need a compression along the "northeast diagonal" of the square element in Figure 4.8 and a tension along the "southeast diagonal." The net effect of 317 MPa must lie between the 310 and the extra *compression*, rather than between the 310 and the extra tension.

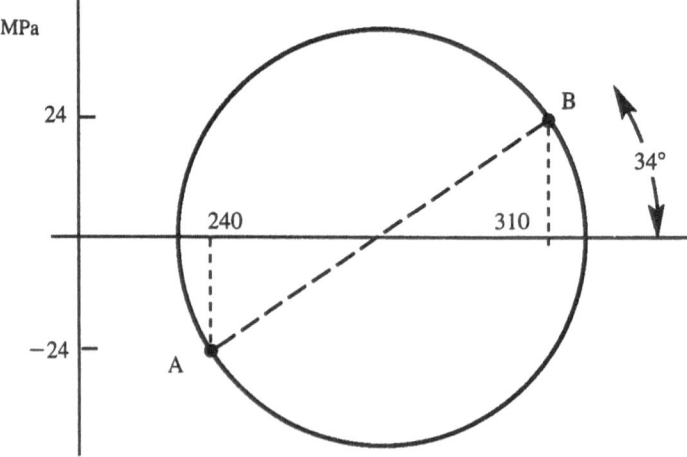

FIGURE 4.22A

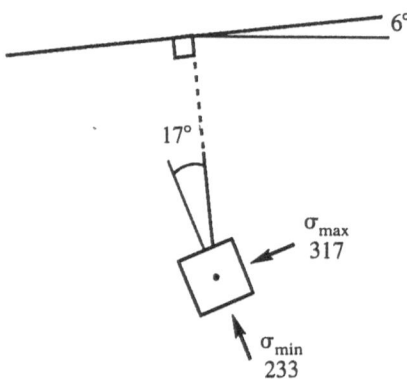

FIGURE 4.22B

Answer 4.10

(a) Shear stress on plane Q is equal in magnitude to the shear stress on plane P—i.e., 40 MPa.

(b) See Figure 4.23.

(c) $2\theta = 70°$; $\theta = 35°$.

(d) See Figure 4.24.

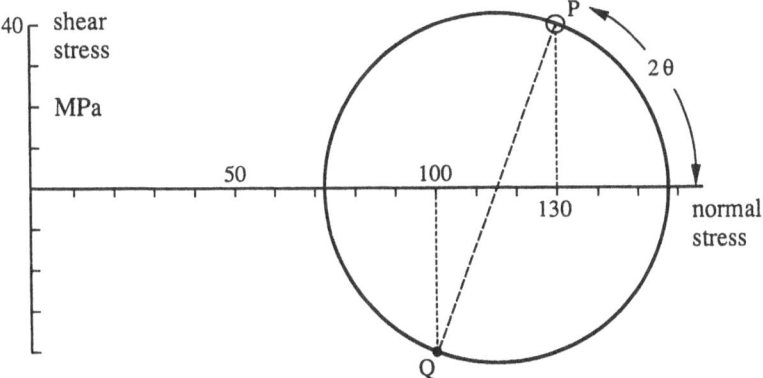

FIGURE 4.23

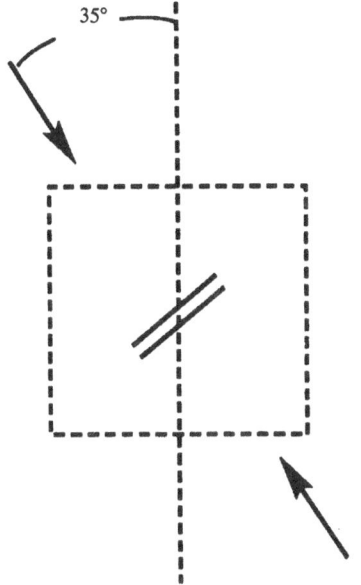

FIGURE 4.24

Answer 4.11

(a) Shear stress on plane *S* is equal in magnitude to the shear stress on plane *R*—i.e., 40 MPa.

(b) See Figure 4.25.

(c) $2\theta = 40°$; $\theta = 20°$

(d) See Figure 4.26.

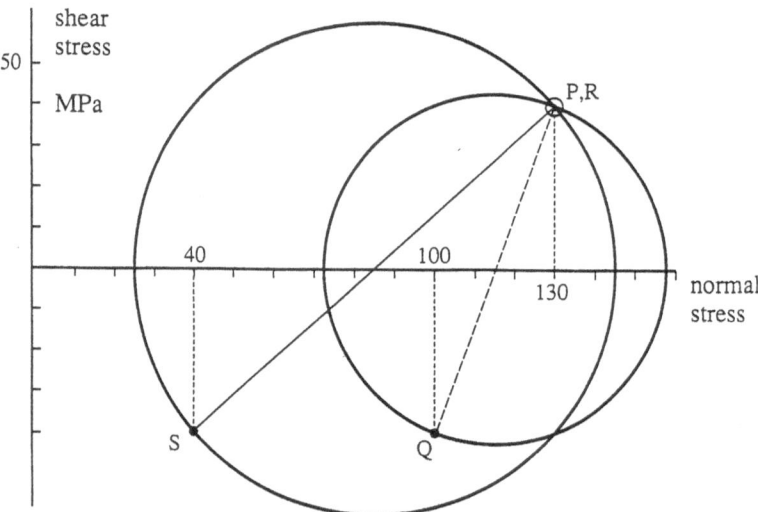

FIGURE 4.25

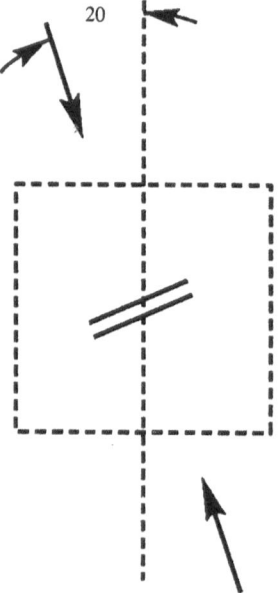

FIGURE 4.26

Answer 4.12

A complete picture of the stress states can be made as in Figure 4.27, with lengths of lines proportional to the magnitudes of the principal stresses. The magnitudes of course are very clearly seen in Figure 4.25 as well, but the orientations are best seen in Figure 4.27. The change in direction of σ_{max} across the interface is called "refraction of stress."

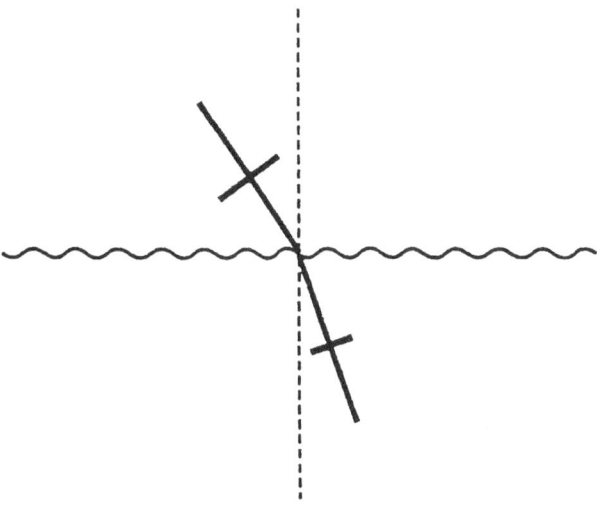

FIGURE 4.27

CHAPTER 5

Rheology: Relations Between Forces and a Material's Response

Familiar ideas are that some materials behave elastically, like rubber, and some behave viscously, like honey and engine oil. In some materials, both kinds of behavior are important simultaneously, and having elastic properties certainly does not prevent a material from having viscous properties as well. To get started on the topics of this chapter, viscous behavior in rocks will be emphasized; the groundwork thus established will be helpful in later discussion of more complicated behaviors.

In **viscous** behavior, energy is dissipated (mostly as heat) rather than being stored, and the material has no tendency to revert to an earlier shape: upon removal of forces, a sample simply sits passively; museum specimens of folded rock layers are good examples. The strains are therefore termed **nonrecoverable,** in distinction from the recoverable strains one sees in elastic behavior.

SHEAR STRESS, SHEAR STRAIN, AND VISCOSITY

To establish the magnitude of a material's viscosity, one has to focus on a plane in the material. Then one may inspect the shear stress and the shear strain rate at this plane; the ratio of the two is the material's **viscosity.** If we designate it by N,

$$N = \frac{\text{shear stress}}{\text{shear strain rate}} \quad \text{sometimes written} \quad \tau/\dot{\gamma}. \tag{5.1}$$

Question 5.1 Viscosity and shear strain rate: If a shear stress of 100 MPa produced a shear strain rate of 0.1 per million years (abbreviated "my"), as might have happened in the deformation of some known fossils, what was the material's viscosity? (The answer may appear as so many MPa-my, and this is a permissible viscosity unit.)

Question 5.2 Viscosity and shear strain rate, continued: A rock with viscosity 30 gvu as in Answer 5.1 sits in a shear zone under a shear stress of 80 MPa. What is its strain rate? If conditions remained steady for 10 my, how much strain would accumulate?

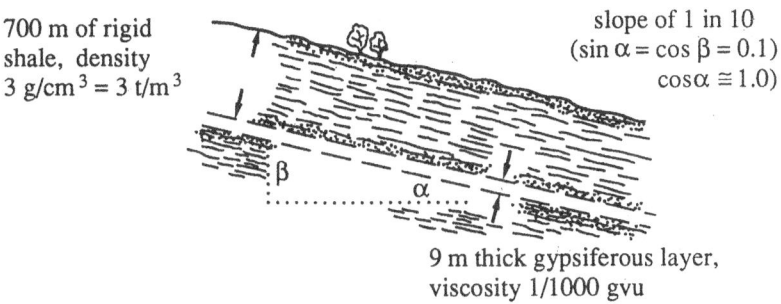

700 m of rigid shale, density 3 g/cm^3 = 3 t/m^3

slope of 1 in 10 ($\sin \alpha = \cos \beta = 0.1$) $\cos \alpha \cong 1.0$)

9 m thick gypsiferous layer, viscosity 1/1000 gvu

FIGURE 5.1

Question 5.3 Viscosity and displacement rate: At what rate does the sheet of shale in Figure 5.1 slide downhill under its own weight?

Question 5.4 Shear of a layered succession: An alternating succession of horizontal shale beds 12 cm thick and limestone beds 8 cm thick is subjected to a steady layer-parallel shear stress of 24 MPa. Shale viscosity = 30 gvu; limestone viscosity = 80 gvu. If an organism bores a vertical burrow, calculate the deformed shape of the burrow 6 2/3 my later and draw it at, say, 1/4 × true size (6 2/3 my = 2 gtu). (This topic is continued from Question 2.27.)

Question 5.5 Weighted mean viscosity: If the organism in Question 5.4 had been hyperactive and penetrated 100 beds vertically in a period of time that was geologically negligible, what would be the subsequent deviation of its burrow from vertical on a gross scale, disregarding the lithologic alternation? Answer in degrees, but also in meters sideways per meter vertically. Hence calculate an overall shear strain rate for the assemblage, and an overall or effective viscosity.

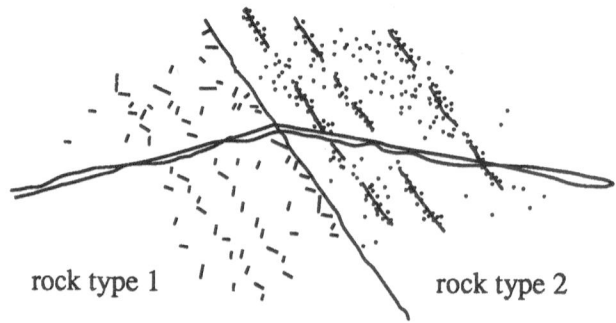

rock type 1   rock type 2

FIGURE 5.2

Question 5.6 Viscosity ratio: What is the ratio of viscosities for the two rocks in Figure 5.2, if the marker was formerly straight and normal to the interface?

The answer to Question 5.6 depends on assuming that (shear stress)$_1$ = (shear stress)$_2$ at all times, which is a safe bet. The calculation remains valid however much the shear stress varies in time (subject to the restriction below) and also remains valid even if the rocks flatten and stretch parallel to the interface. The rocks can even dissolve away without affecting the validity of the answer, as long as they do so to the same extent (in fractions or percents) on both sides of the interface. But if the limestone thinned by dissolution while the shale did not, the marker could become misleading. Misleading markers cause recurrent difficulty in structural geology, but they don't cause anything like as much difficulty as absence of markers.

The restriction mentioned is as follows. Change of shear stress with time does not in itself affect the validity of the calculation indicated, as long as each rock's viscosity *remains constant*. However, if the history involves change of stress and as stress changes, the rocks' viscosities change also, the ratio could be less than 2.5 at some stage and more than 2.5 at another. The extent to which a rock's viscosity might change as the stress magnitude changes is discussed later in this chapter, in the section titled *A Rock Has Variable Viscosity*.

Nobody has good information at present about how large a source of error this might be.

Question 5.7 Shear strain in a fault zone: Suppose that the Altyn Tagh Fault gives way when the shear stress on it reaches 40 MPa. In a quiet interval between earthquakes that lasts 1/1000 gtu (3300 years), what is the *most* shear strain that could accumulate

(a) in the fault zone, where effective viscosity of the rock = 0.08 gvu?
(b) in rock 30 km away from the fault zone, where effective viscosity = 50 gvu?

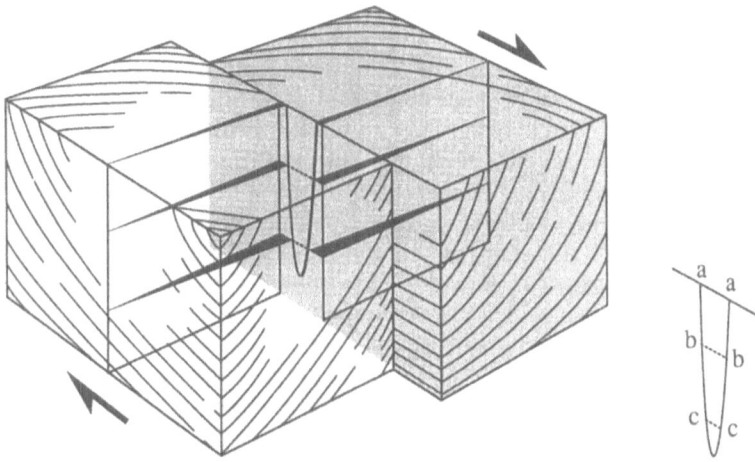

FIGURE 5.3 The diagram shows two blocks of lithosphere displaced with respect to each other by about 10 km (total depth shown is about 30 km). The total offset is the same at all levels, but the manner in which the offset is achieved varies: at the surface, two horizontal lines initially normal to the fault plane get *slightly* curved and their tips become separated by nearly 10 km of seismic slip (aa). At lower levels, the amount of displacement taken up by curvature increases, and the amount of slip decreases (bb and cc). The diagram goes down to a level where no slip occurs, all the displacement being accommodated by continuous processes.

Question 5.8 Shear strain in a fault zone, continued: Consider two lithospheric plates that move in a strike-slip manner at 3 cm/year, or 100 km/gtu. At, say, 30 km depth, the displacement is probably spread over a wide zone and achieved entirely by steady creep with no slip (no jerks or earthquakes). At a shallower depth, part of the 100 km is accommodated by creep, perhaps 60 km, and the remaining 40 km occurs by slip, e.g., 33,000 jerks of just over 1 m each, at an average rate of 1 per 100 years. At a shallower depth still, perhaps only 10 km of the total is accommodated by creep (see Figure 5.3).

(a) What shear-strain rates do we envisage, in order of magnitude?
(b) If the shear stress at the surface oscillates between 10 MPa and 50 MPa (with stress drop of 40 MPa at each earthquake, and a mean value over time of 30 MPa), what kind of stress magnitude and viscosity value could give the shear strain rate offered in (a) at 30 km depth?

To get started on Question 5.8(a), one has to speculate whether the shear strain is distributed over a width of 20 km or 200 km, or somewhere between the two. Be bold: choose some possible value to work with.

Is "Plate Tectonics" True? The parallelism of magnetic stripes, and the geometrical fit of South America into Africa suggest that plates retain their surface dimensions, over periods such as 200 my, with changes no greater than a few kilometers. What can we conclude about shear stresses and viscosities *away from* the mobile zones or active marginal parts of the lithosphere? Use as a working value the idea that, in map view or as seen by satellite, plates suffer less than 3° of shear strain in 200 my.

$$3° = 1/20 \text{ in shear strain units, and } 200 \text{ my} = 60 \text{ gtu;}$$

therefore the maximum shear strain rate envisaged = 1/1200 per gtu. For example, suppose average shear stress = 1 MPa; then minimum viscosity = 1200 gvu. This is higher than most viscosity values proposed in this book, in conformity with the idea of stable cratons and mobile belts. The almost-rigid dead interior of a continent may indeed have an effective viscosity greater by 100 or 1000 times than the viscosity of its more mobile margins. Possible reasons for the contrast are touched on at the end of Chapter 7.

A second comparison with ideas from plate tectonics can be made using Answer 3.29. It was there speculated, for the purpose of the exercise, that there might be an east-west shear stress at the base of the lithosphere under the South Atlantic of 6 MPa. Could such a stress reasonably arise from the asthenosphere dragging on the lithosphere?

To answer the question, we need an estimate of the strain rate in the asthenosphere—a very poorly known quantity. Take an average mid-ocean spreading rate as 5 cm/yr or 50 mm/yr or 50 km/my or 170 km/gtu; then let us suppose that the asthenosphere moves twice as fast as a lithosphere plate at a depth 200 km below the plate. This would involve the asthenosphere moving with respect to the lithosphere at 170 km/gtu, and straining at a shear strain rate of 170/200 or 0.85 per gtu. If this shear strain rate were in fact accompanied by a shear stress magnitude of 6 MPa, the material's viscosity would be 6/(0.85) or about 7 gvu.

The preceding calculation is highly speculative. For all we know, the situation may be reversed, with the lithosphere being the active agent that drags on the asthenosphere. But there is nothing wrong with the basic concepts: forces must balance, and strain rates must relate to stresses through the material's viscosities. Whether we look at fossils and worm tubes or continents and oceans, we can continue to have confidence in the same principles.

Summary and Comment

Shearing processes are common throughout geology, and the idea of a rock having a viscosity is useful for making estimates and predictions. However, rocks are far more complicated than the simple fluids, such as water and oil, from which the idea of viscosity arose. We shall shortly consider the question: Does a rock actually flow, in the same sense that water flows? but before that discussion, we shall forge on and apply the idea anyway, to a second group of examples where normal stresses and linear strains are the quantities of interest.

Relation of GTU and GVU to Standard Units

The main purpose for using the units "gtu" and "gvu" is to give a sense of scale for metamorphic rocks—to permit calculations about deep-seated processes of structural geology without constantly manipulating large powers of 10. However, it is necessary to be able to convert to the International System of Units, SI units (kg, m, s), and it is also necessary to keep in mind the *large* range of times, stresses, and viscosities we sometimes need to range over: though an orogeny may last for tens of gtu, single events that belong to it may be *far* more brief; strain rates range from less than 1 per gtu to 1 per second. The following equivalences are therefore sometimes needed; see also the inside back cover.

Time: 1 gtu $= 10^{14}$ sec $\cong 3^1/_3$ my
              1 my $\cong 3 \cdot 10^{13}$ sec
              1 year $\cong 3 \cdot 10^7$ sec   (more exactly, $3.157 \cdot 10^7$)

Viscosity: 1 gvu $= 1$ MPa-gtu $= 10^{14}$ MPa-sec
                    $= 10^{15}$ bar-sec $= 10^{20}$ Pa-sec
                    $= 10^{21}$ poise

           1 MPa-sec $= 10$ bar-sec $= 10^6$ Pa-sec $= 10^7$ poise

           1 bar-sec $= 10^5$ Pa-sec $= 10^6$ poise

NORMAL STRESSES AND LINEAR STRAINS

Opportunity exists for making this topic a good deal more complicated than the topic of shear stress and shear strain. However, in certain circumstances, matters continue on the same simple lines, and many geological questions can be explored and interesting ideas checked out. The circumstances are

1. no change in a sample's volume, and
2. no change in dimensions perpendicular to the plane observed.

These combine to establish, for the plane observed,

   maximum shortening strain rate $= -$(maximum elongation strain rate).

In this circumstance,

$$\text{Maximum shortening strain rate} = \frac{\text{stress difference}}{4\,N}, \qquad (5.2)$$

which is often written

$$(\sigma_{max} - \sigma_{min})/4N,$$

where σ stands for a compressive normal stress. The reason for the factor of 4 can be seen by looking at the Mohr circles for stress and strain rate, Figure 5.4.

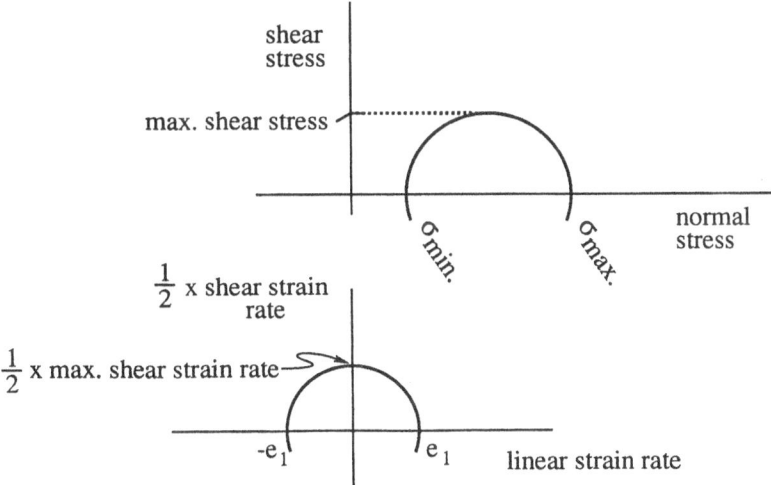

FIGURE 5.4   Maximum shear strain rate $=$ shear stress$/N$, so radius of strain-rate circle $=$ radius of stress circle$/2N$, $=$ diameter of stress circle$/4N$.

Question 5.9 The Pyrenees: In Answer 2.11, a shortening strain rate of 1/6 per gtu was estimated for a horizontal north-south line during formation of the Pyrenees. Assume that this was accompanied by an elongation strain rate of 1/6 per gtu vertically so that the formula just introduced can be used. If the driving stress difference was 80 MPa, what was the effective viscosity of the deforming mass?

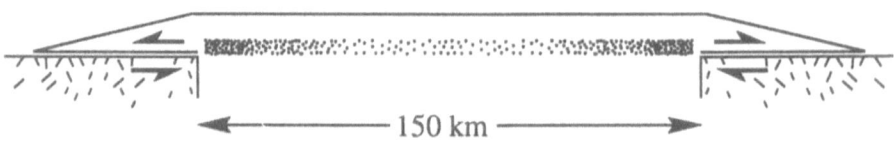

FIGURE 5.5

— 150 km —

Question 5.10 The Pyrenees, continued: Assume that the overburden presses on the stippled portion of Figure 5.5 with a uniform stress of 120 MPa vertically, and that there is a horizontal restraint of 84 MPa on each end.

(a) At viscosity 120 gvu, what linear strain rate do you expect in the stippled portion?
(b) Assuming its centerpoint stays still, at what rate will its southern end move southward?

Questions 5.9 and 5.10 draw attention to the abrupt transition that probably exists during jaw-closing events, between horizontal shortening at lower levels and horizontal elongation slightly higher up. As noted in Question 3.32, there are also vertical forces to be balanced. Any reasonable hypothesis for the Pyrenees needs to weave these three subproblems together in a compatible way. (The topic is continued in Chapter 7.)

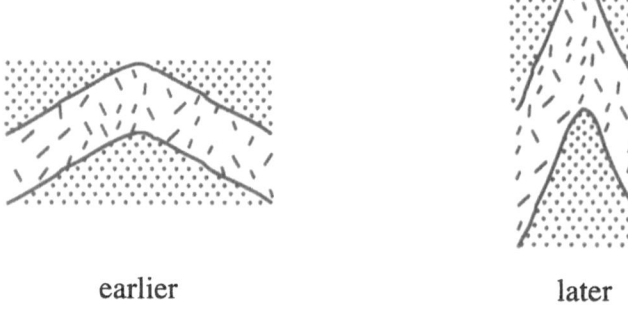

earlier   later

FIGURE 5.6

Question 5.11 Passive folding: The conspicuous mountains of the present earth formed during the Tertiary period—i.e., during the last 70 my or the last 20 gtu. Any single fold must have formed in a shorter time such as 2 gtu or possibly much less. Let us consider possible stress magnitudes and viscosities for the passive fold in Figure 5.6. The horizontal shortening is 1/2 × original width, and the vertical elongation is 1 × original height, and if this change takes 2 gtu, the linear strain rate must be, in order of magnitude, 1/4 to 1/2 per gtu; or if the fold develops in only 0.2 gtu (1/100 of Tertiary time), a typical linear strain rate would be 2 1/2 to 5 per gtu. Calculate strain rates for the 12 boxes shown in Table 5.1 and see which ones fall in the right range for the passive folding discussed.

TABLE 5.1

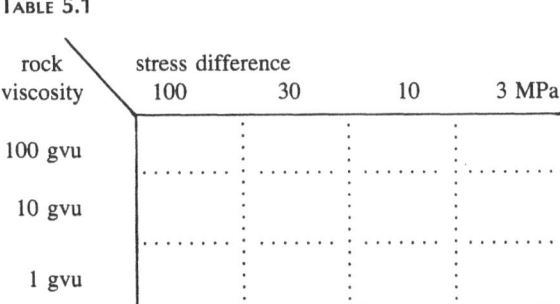

rock viscosity	stress difference 100	30	10	3 MPa
100 gvu				
10 gvu				
1 gvu				

The strain rates calculated in Answer 5.11 suggest that rocks in course of folding have viscosities 100 gvu or less. Stress differences greater than 100 MPa crush rocks at shallow depths rather than folding them, so that figure is an upper limit for shallow depths and, by an indirect argument, a reasonable working upper limit at greater depths also (more extended discussion in Chapter 7). Hence the rock viscosity must have dropped below 100 gvu at the time of folding.

It is less easy to establish a minimum value. In unconsolidated sediments, we might form a fold in 1 hour under 0.1 MPa, with a viscosity that is tiny compared with 1 gvu. Exactly how much viscosity a shale loses during those times when its contained chlorite is dehydrating (the whole mass being hot and unstable) and how long such a low-viscosity state might last are very significant unknowns.

Question 5.12 Linear strain in a shear zone: Reconsider the shear zone, Question 5.2, in terms of compressive stresses and linear strain rates. Assume the shear stress of 80 MPa is the *maximum* shear stress present. What linear strain rates will result?

The directions of σ_{max}, σ_{min}, $\dot{e}_{max}$ and $\dot{e}_{min}$ are at 45° to the plane of τ_{max} and $\dot{\gamma}_{max}$. The fact that the material and the stress field are rotating with respect to each other makes no difference to the relations between stress and strain rate, and the mean value $(1/2)(\sigma_{max} + \sigma_{min})$ makes no difference to the relations either; but if we transferred to a new value of the mean stress, we might find that the rock behaved with a viscosity no longer equal to 30 gvu.

Question 5.13 Boudinage: An alternating succession of horizontal shale beds 12 cm thick and limestone beds 8 cm thick is subjected to a vertical compression of 60 MPa. The horizontal compressive stress in the shale layers is 40 MPa. What horizontal compressive stress in the limestone layers will permit them to flatten and elongate at the same rate as the shale? Continue to assume conditions (1) and (2) from page 117 for conservation of area. Viscosities are, for shale, 30 gvu and, for limestone, 80 gvu, as in Question 5.4.

The results from Question 5.13 are shown in Figure 5.7. The filled circle shows σ_{horiz} in the limestone, the open circle shows σ_{horiz} in the shale, and any other horizontal line will cut the two sloping lines at points such that

$$(\text{stress difference})_{\text{limestone}} = 8/3 \times (\text{stress difference})_{\text{shale}}$$

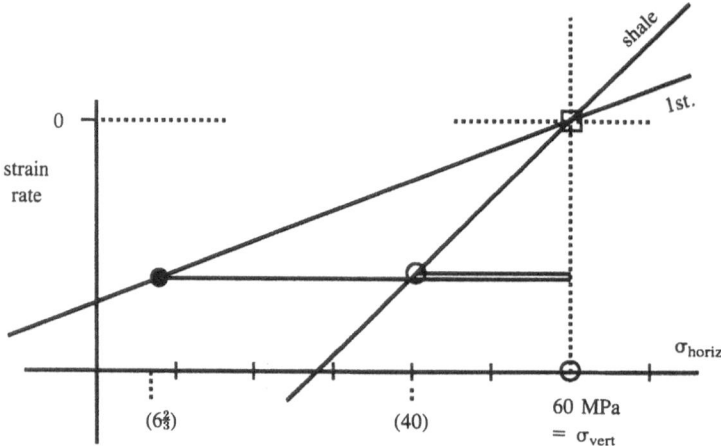

FIGURE 5.7

so that

$$(\text{elongation strain rate})_{\text{limestone}} = (\text{elongation strain rate})_{\text{shale}}.$$

EXAMPLE: suppose $(\text{stress difference})_{\text{limestone}} = 8$; then $(\text{stress difference})_{\text{shale}} = 3$ and joint strain rate = 1/40 elongation per gtu. Horizontal compressive stresses would be 52 MPa and 57 MPa in limestone and shale respectively.

Question 5.14 Alternating shale and limestone, shortening: What stress states would give horizontal shortenings of 1/40 per gtu in the limestone and in the shale, still with 60 MPa vertical stress?

Question 5.15 Alternating shale and limestone, continued: What strain rate and what stress difference in the limestone go with $(\text{stress difference})_{\text{shale}} = 9$ MPa?

Two questions are of interest:

1. What about the possibility of the shale and limestone flattening and elongating at different rates? (*Optional:* Readers interested in this question, go to page 121.)
2. What happens if we try to achieve a flattening rate of say, 0.2 per gtu, requiring a stress difference of 24 MPa in the shale beds (and horizontal compression of 36 MPa)?

For the answer, we need a stress difference of 64 MPa in the limestone, which could only be got by having, horizontally, a *tensile* stress of 4 MPa. One might say, "Aha! we have now encountered a condition that might cause the limestone bed to part into boudins," but that would be taking an oversimple view. One needs to remember (i) the presence of fluids, (ii) the fact that fluid in the shale is affected by the stress state in the shale, and (iii) the fact that if fluid in the shale gets to have a pressure greater than the least compressive stress in the limestone, fluid is likely to migrate into the limestone and open up fissures there.

For example, suppose p_{fluid} in shale $= 0.6 \times$ mean stress in shale. Then for the first condition calculated, and shown by the circles in Figure 5.7, $p_{fluid} = 30$ MPa, far more than enough to open vertical fissures in the limestone if the horizontal compression were only 6 2/3 MPa.

For more on the ability of fluids to create fractures, see Chapter 6. For migration of fluids without fracturing, see the last section of this chapter. The topic of boudinage in particular is continued at Question 6.14.

Shale and Limestone Layers Spreading at Different Rates   The calculations in Answers 5.13 through 5.15 emphasized the idea of alternating layers spreading at the same rate, but we might question whether this is realistic. If the shale has a lower viscosity, will it not spread at a faster rate? Yes, it will, and the situation in Question 3.30 will develop, at least to some extent. Even so, the calculation in Answer 5.13 is of value: the stress magnitude of 6 2/3 MPa is the *limiting value* in the center of the limestone slab. As long as parting does not occur and foul up the whole stress system, the horizontal stress in the limestone will drop closer and closer to 6 2/3 MPa, but will never drop below this value. The reason is that the drag exerted by the shale arises from the shale's spreading faster: the closer $\sigma_{horiz,limestone}$ comes to 6 2/3 MPa, the closer the limestone comes to spreading as fast as the shale and the less the drag of one on the other. For broad spreads of an alternating set, there will be complications at the margin, wherever that is, but through much of the center, the condition of equal spreading rates will be *closely approached*, even though never exactly attained. Hence, calculations using that simple idea are a powerful, fairly reliable indicator.

Coherent and Incoherent Interfaces   An important idea emerges if we consider the shale–limestone interface in a little more detail. Suppose that when the 60 MPa load normal to layering is first imposed, the compressive stress parallel to layering is 26 2/3 MPa uniformly through all layers. The stress difference will be 33 1/3 MPa everywhere, and spreading strain rates would be 0.28 and 0.10 if the shale and limestone did not interact at all (e.g., if their surfaces were perfectly lubricated). But in reality they will interact; lines initially straight and vertical in the shale will become barrel-shaped; because the barrel effect increases outward toward the region into which the layers are spreading, the shear stress at the interface or "drag stress" will also increase outward, as in Figure 3.17. Near the center, this stress will be less than the interface can withstand, but farther out it might become greater. Thus we might find the situation shown in Figure 5.8, where in the central region the limestone and shale do remain coherent, with marker lines remaining continuous (though distorted). Farther out, the interface parts, allowing slip; markers become discontinuous and the interface becomes incoherent.

Whether the situation tends toward the condition already calculated, with compressive stress parallel to the layering rising to 40 MPa in the shale and dropping to 6 2/3 MPa in the limestone, depends on conditions outside the left and right margins of Figure 5.8. Under some conditions, the compressive stresses might continue mostly close to 26 2/3 MPa. But where the interface is coherent, the drag of the shale must locally reduce the compression in the limestone to 6 2/3 MPa, and the drag of the limestone must locally increase the compression in the shale to 40 MPa—or, if not these values, some other pair of values from Figure 5.7 must develop, because as long as the interface is coherent, the strain rates on the two sides must at least locally be equal. (Away from the interface, at the mid-level of any layer, the strain rates can differ.)

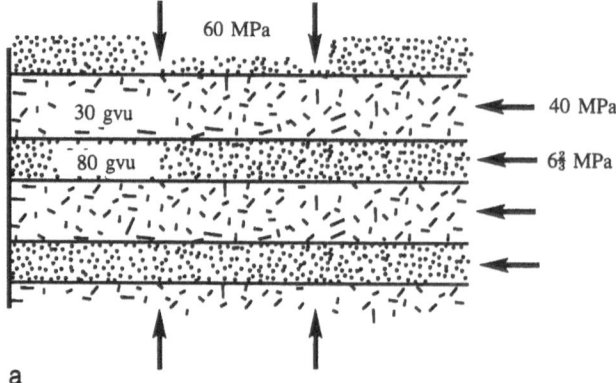

FIGURE 5.8A

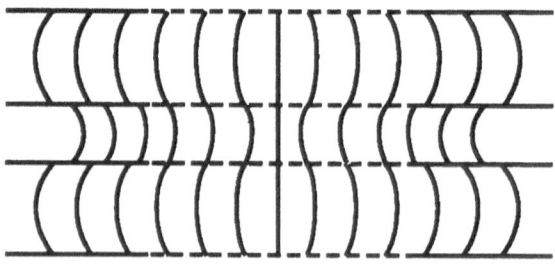

b

FIGURE 5.8B

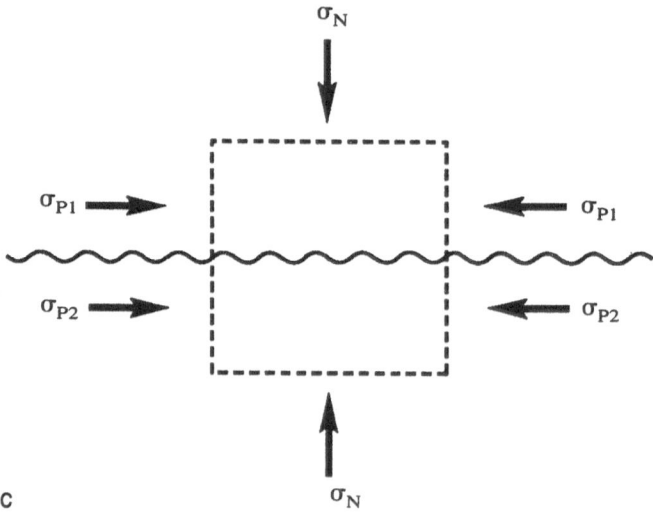

c

FIGURE 5.8C

If the interface between two materials is coherent, then immediately adjacent to the interface the stress differences in the two materials must be in the same ratio as the materials' viscosities. Here "stress differences" means $\sigma_N - \sigma_P$, where σ_N and σ_P are shown in Figure 5.8(c).

$$\frac{\sigma_N - \sigma_{P_1}}{\text{viscosity 1}} = \frac{\sigma_N - \sigma_{P_2}}{\text{viscosity 2}} \qquad (5.3)$$

Summary and Comment

For linear strains, as for shear strains, it is clearly possible to have *some* success describing geological deformations in terms of steady stress magnitudes, uniform strain rates, and constant viscosities. However, reference has already been made to the probability that all three quantities fluctuate; no calculations of the type so far made should be taken seriously as descriptions of what actually happened. The calculations are a useful check on orders of magnitude, or a useful preliminary to more realistic (and more complicated) speculations. Two prominent complications are introduced in the next section.

A ROCK HAS VARIABLE VISCOSITY

The mechanisms by which a rock deforms operate on the scale of a single crystal or smaller scale—mountains are built by the movement of grains. Much of the action occurs in the small regions between one grain and the next. Examples of these microprocesses are

grain-boundary slip, of one grain past its neighbor;
solution of material at some points and deposition elsewhere;
bending and cracking of grains, until the fragments separate and take on independent lives.

These processes interact; any one operating singly would be rather ineffective. For example, grain-boundary slip is hindered when a projection meets a barrier, and goes faster again after the obstruction either cracks off or dissolves away. A second feature is that the ensemble of processes is sensitive to fluids: a small change in the amount or pressure of the pore fluid can produce a large change in the overall mobility of the rock. A third feature is feedback: deforming a rock affects its deformability. If we wish to continue summarizing a rock's behavior by the viscosity relations in the preceding sections, to allow that the viscosity changes during deformation is essential.

Question 5.16 Variable viscosity: Refer to the 9 m gypsiferous layer in figure 5.1, sliding 0.6 cm/yr under 700 m of shale. Suppose that when the shear strain reaches 0.3, the viscosity drops to one-third of its former value, and when shear strain reaches 0.6, the viscosity drops to one-third again (one-ninth of original). How long does it take to accumulate the first 0.3 of strain? and the second? and the third?

The effect of strain-dependent viscosity is rather striking where the rock is initially a little inhomogeneous. Suppose the 9 m gypsiferous layer were replaced by a sandwich in which the central 3 m had initially a viscosity half as great as the upper and lower thirds, and was affected by strain in the same way. What is the history of sliding now?

Using numbers from Questions 5.16, we figure that in 225 years, the central layer will reach a strain of 0.3, while the upper and lower are still at strain 0.15. After 75 years more, the central layer reaches strain 0.6 while the upper and lower reach 0.2. After 25 years more, the central layer reached 0.9, and so on. The effect of strain on viscosity is the same in all layers, and the initial variation is only by a factor of 1/2, but later the contrast is much more because of the feedback. Again the overall effect is realistic: if a thick layer of material begins to give way, commonly the later action gets to be concentrated in small regions of high strain. In reverse, one might say that any episode that involves fracture or parting toward the end involves this type of self-accelerating diminution of viscosity in the preliminary stages. When a rock gives way, it is not only because the stress has increased; it is also because the rock's ability to resist has decreased. (The topic is continued in Chapter 7 under the heading A Changing Population of Microfractures.)

Variation in Shear Stress

The formula

$$\text{shear strain rate} = \text{shear stress/viscosity}$$

leads toward the statement

$$\text{shear strain} = (\text{shear stress/viscosity}) \times \text{time}.$$

The latter is of course true for any interval of time during which both shear stress and viscosity are constant, but not otherwise. The preceding section emphasized variation in viscosity; it is just as important to keep in mind variation in shear stress.

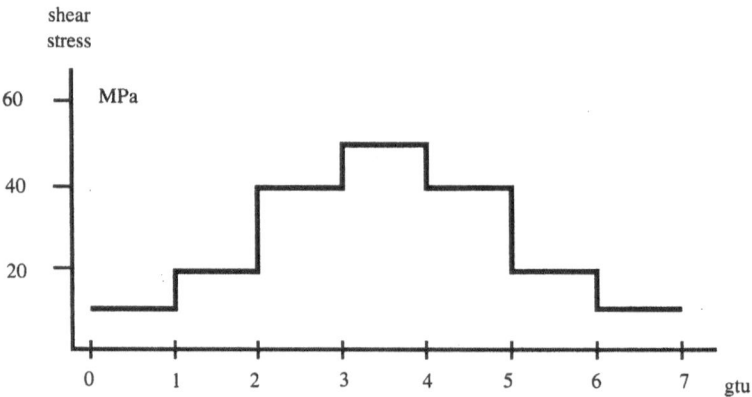

FIGURE 5.9

Question 5.17 A transient stress pulse: Suppose that for a deformation event, the stress history can be approximated by the magnitudes in Figure 5.9. For a horizontal layer of rock with viscosity 50 gvu, show the attitude of an initially vertical worm tube at each stage.

In Answer 5.17, displacement arrows give a better impression of the high activity during the fourth gtu than the successive positions. The predominance of the fourth gtu is even greater if we allow for the fact that, in many conditions, rocks seem less viscous at higher stress. Suppose that the stress history is as in Question 5.17 above, but the rock's viscosity varies with the stress magnitude thus:

stress	10	20	30	40	50	MPa
viscosity	60	30	20	15	12	gvu.

Then the strain increments in each of the seven gtu's are

0.17	0.67	2.7	4.2	2.7	0.67	0.17

and the cumulative strains are

0.17	0.84	3.5	7.7	10.4	11.1	11.3

as in Figure 5.10.

If viscosity diminishes as stress climbs up, the action becomes relatively more concentrated in time. In numbers, for the exercise just shown, the peak stress is 1.8 × mean stress but the peak strain rate is 2.6 × mean strain rate.

For geological realism, we should combine the effects of strain and stress. In the preceding exercise, the seven viscosity magnitudes used were as shown in Table 5.2 at line 3. The strain increments on line 5 lead, as before, to a cumulative strain of 0.84 after 2 gtu. At this point, and *because of this strain,* we enter gtu 3 with a viscosity that is shown on line 4, that has been somewhat arbitrarily diminished below the value on line 3

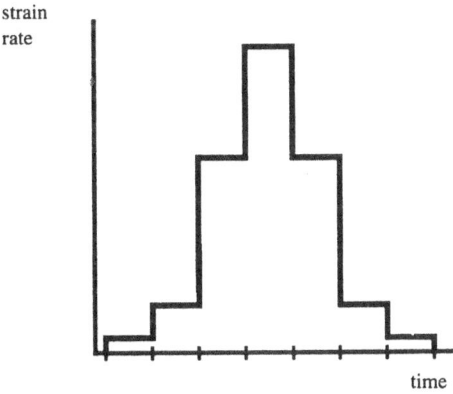

a

FIGURE 5.10A

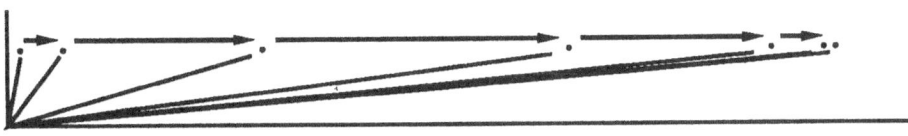

b

FIGURE 5.10B

TABLE 5.2

1	gtu		1	2	3	4	5	6	7
2	stress		10	20	40	50	40	20	10 MPa
3	viscosity		60	30	15	12	15	30	60 gvu
4	viscosity diminished by strain		—	—	10	6	5	8	15 gvu
5	strain increment	0.17	0.67	4.0	8.3	8.0	2.5	0.7	
6	cum. strain		0.84	4.84	13.2	21.2	23.7	24.4	

to allow for the effect of the strain. For purposes of the example, round-number values have been entered along line 4, each being a successively smaller *fraction* of line 3 (but with the values rising again toward the end as stress decays). The strain increments are shown in Figure 5.11.

Consequences are, first, that the strain increments and strain rates are larger than before and, second, that the strain-history curve becomes offset from the stress-history curve. Both conclusions are geologically realistic, but just as important is the idea the example illustrates: viscosity continues to be the rock property we need for speculating about deformation history, but the viscosity value envisaged at any moment needs to be appropriate to that moment, being affected by the rock's current textural state as well as by the current stress level.

In Table 5.2, the viscosity values range from 5 gvu to 60 gvu, and the strain rates range from 0.17 to 8.3 per gtu. Despite the variations, one may, if one wishes, calculate

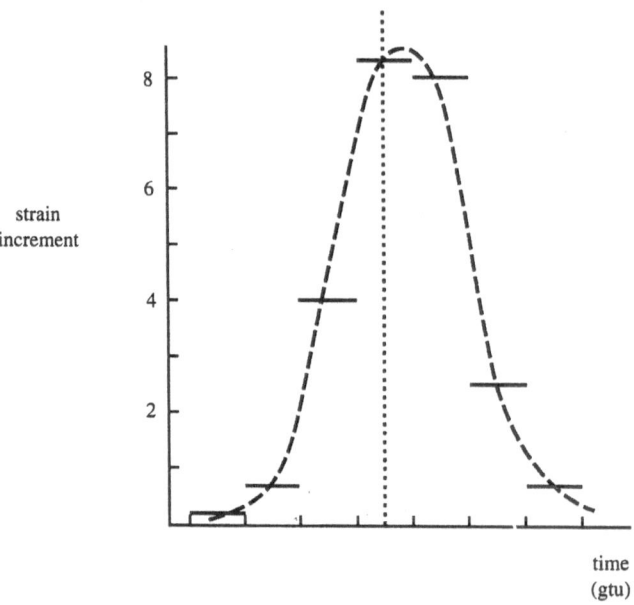

FIGURE 5.11

averages. Thus, average strain rate $= (24.4)/7 = 3.5$ per gtu, and average stress $= 27$ MPa. Hence, for the entire deformation episode, the average viscosity of the rock was 7.7 gvu, and it is true that

$$\text{total strain} = \text{time interval} \times \frac{\text{average stress}}{\text{average viscosity}}.$$

When we perform calculations using this one-line statement with average values, as we did in Answers 5.2 through 5.5, the results are valuable guides as to order of magnitude. But it is helpful to keep in mind the way a rock's viscosity can vary and the largeness of the viscosity's range. During deformation, a rock has an evolving suite of instantaneous viscosity values, most of which are less than its viscosity in its long-term state, or what one is tempted to call its "undisturbed state" as line 4 of Table 5.2 illustrates.

Other important factors that influence a rock's viscosity are temperature, total compressive stress, and behavior of the pore fluid. Once the idea is entertained of a rock's viscosity being continually updated, and always reflecting the conditions of the moment—including the rock's current texture—there is no new principle involved in allowing for these factors. (The topic is continued in Chapter 7, under the heading Effect of Pore Fluid on Flow Behavior.)

Summary   The exercises dealt with a period of 7 gtu, or about 23 my, within which there was a shorter period during which most of the deformation occurred. The exercises were introduced by reference to a single horizontal layer with worm tubes, but the time scale is perhaps more appropriate to the movement of a thrust sheet or lower limb of a nappe. The total strain in Table 5.2 is 24.4 and this fits, for example, a lower limb 400 m thick with 10 km of displacement accommodated on it, as in Figure 5.12. Or some portion of the thrusting in the southern Pyrenees, Figure 5.5, might be a real-life instance of total time and total strain of these orders of magnitude.

PRINCIPAL STRESS INCLINED TO AN INTERFACE

Shear zones, limbs of folds, and kink bands are all features that can be understood only by imagining principal stresses inclined to an interface, or layer, or set of layers. At the margin of a rising dome, again, the lithologic interface and the stress field will make some not particularly simple angle. Such problems can be handled by techniques already covered, as long as the area-conserving assumptions are permissible as set out on page 117. In brief, the stress field is broken up into components parallel and perpendicular to the interface; then the shear stress parallel to the interface is handled as on pages 113ff., while the normal stresses parallel and perpendicular to the interface are handled as on

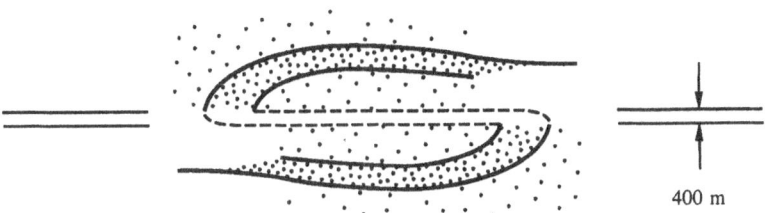

400 m

FIGURE 5.12

pages 117ff. No new equations are involved; we simply combine ideas that we have used before, imagining the processes to be going on together instead of going on at separate times or places.

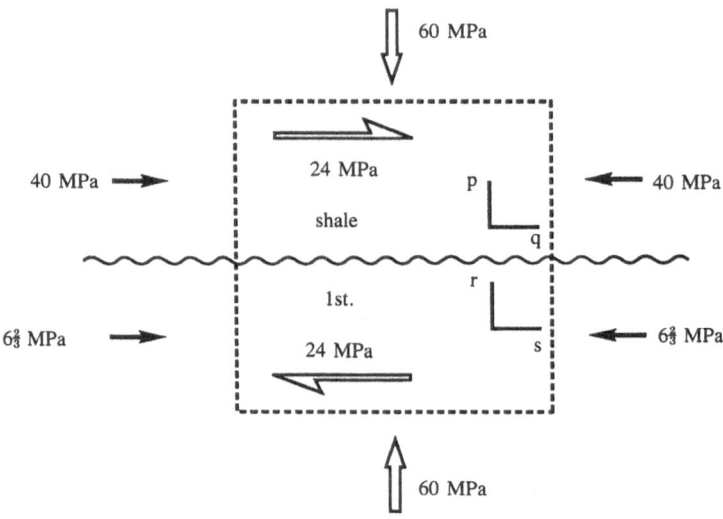

60 MPa

40 MPa →

24 MPa

shale

p

q

← 40 MPa

r

1st.

24 MPa

s

6⅔ MPa →

← 6⅔ MPa

60 MPa

FIGURE 5.13

Question 5.18 Strains at an interface: At a planar interface between a limestone unit and a shale unit, stresses are present as in Figure 5.13. Assume that in the plane of the diagram, areas are conserved; viscosity of shale = 30 gvu and viscosity of lime-stone = 80 gvu. What is the linear strain rate along line *p?* line *q?* line *r?* line *s?* What is the shear strain rate of the pair *pq?* What is the shear strain rate of the pair *rs?*

(Questions 5.4 and 5.13 are related.)

We can use the example shown in Figure 5.13 to verify an assertion that was stated without any back-up following Question 5.6. The assertion was made that, for a bent marker as in Figure 5.2, the ratio of viscosities can be deduced from the tangents of the two apparent shear angles, even if these angles have been affected by flattening normal to the interface. We test this assertion by supposing that, instead of being present simultaneously, the shear stresses and normal stresses in Figure 5.13 act sequentially, for 1 gtu each.

Question 5.19 Shear strain followed by linear strain: At the shale–limestone contact in Figure 5.13, if the shear stress of 24 MPa acts by itself as shown for 1 gtu, the shear angles in the two materials will have tangents 0.8 and 0.3. Call these angles α and β. If during the following 1 gtu, the normal stresses act by themselves (with no shear stress), by how much does tan α change? By how much does tan β change? By how much does the ratio tan α/tan β change?

Following the specific example in Question 5.19, the reader will probably agree that, given a ratio tan α/tan β, it does not matter what linear strain we apply normal to the interface nor at what time during deformation we apply it: as long as the strain is the *same in both materials,* tan β will be changed by the same factor as tan α, and the ratio will be left unchanged. The same is true for linear strains parallel to the interface.

The conclusion is (as already stated following Question 5.6) that a pair of shear angles is a very robust indicator for viscosity ratio. The only way to change the angles so as to give a misleading estimate of viscosity ratio would be to introduce linear strain in one material that is *not* present in the other. The interface would then necessarily become *incoherent* as in Figure 5.8. The conclusion is that if an interface remains coherent, shear angles at the interface must have tangents in the same ratio as the materials' viscosities. (Exception: There is a highly specialized circumstance in which the conclusions just stated are not true. If linear strain occurs exactly normal to the interface by solution processes as in Question 2.2 or by introduction of parallel sheets of new material as in Question 2.8, the strain can be different in the two materials without coherence at the interface being disturbed.)

When interfaces are being inspected for the purpose of estimating viscosity ratios, a far more common difficulty or source of error arises in estimating the shear angles. For example in Figure 5.2, if it were known that the marker was formerly at 90° to the interface, progress would be easy and confidence would be high. More realistically, we can only guess the original angle between the marker and the interface; indeed, it is only a guess that the marker was initially straight. It is a rare circumstance that the geometry of a group of markers across an interface can be analyzed so as to give a picture of their original disposition and of the strains they have suffered subsequently. More commonly the situation is only that *if* one is willing to assume *xxx* . . . , one can conclude that *yyy* . . . has occurred; this type of weak conclusion is often the best that can be reached.

Refraction

The idea of refraction arises in connection with

stresses,
strain rates at some instant,
finite or accumulated strains,
markers, such as veins that cross the interface, and
cleavage.

Refraction of stress was described in Chapter 4, and refraction of markers was described earlier in this chapter, especially in Figure 5.2 and Answer 5.4. Where a marker is initially at 90° to the interface, its refraction is linked directly to the shear angles in the two materials, and shear angles are a part of the more complete discussion in Chapter 2. In particular, Figure 2.29 shows the behavior, through time, of three directions:

the direction of greatest instantaneous strain rate, $\dot{e}_{max}$;
the direction of greatest accumulated linear strain, e_{max};
the direction of a marker that was initially normal to the lamination.

If two materials meet at an interface and suffer shear strain parallel to the interface, all three directions will be refracted. The first, $\dot{e}_{max}$, is normally parallel to the direction of minimum normal stress, σ_{min}, and is refracted in exactly the same was as σ_{min} (*Exception:* If the materials are anisotropic, this parallelism is lost; anisotropic materials are discussed in the next section.) The third direction has already been discussed. The second, e_{max}, definitely gets refracted, but the angles of its refraction cannot be specified in a few words and are better shown in an example.

Question 5.20 The line that shows greatest elongation: If the stress system in Figure 5.13 acts continuously for 1 gtu, what will be the directions in the two materials of the line of greatest elongation, e_{max}? (For answer, use a graphical method based on the picture of the deformation, Figure 5.38 in Answer 5.18. In this diagram, the original shape of the element shown is a square.)

The manner of refraction of e_{max} is very different from that of an original 90° marker such as the line *tru*. It is also different from the refraction of σ_{min} and $\dot{e}_{max}$, which can be found by the method in Answers 4.10 through 4.12.

Question 5.21 Refraction of stress and strain rate: Find the directions of σ_{max}, σ_{min}, $\dot{e}_{max}$, and $\dot{e}_{min}$ for both materials in Figure 5.13.

Observables at an Interface   A contrast is apparent: given a stress state, the simple way to calculate its consequences is to work with the components parallel and normal to the interface, as in Answer 5.18; angular relations are not needed if the objective is to describe how the material responds to known stresses. On the other hand, outcrops more often provide information about *angles* than about linear strains (and hardly ever provide information about stresses); it is when we try to reconstruct stress states from observables that we need to understand the angular relationships. The basic idea of the present text is that one should *first* understand the mechanics of material behavior and *then* embark on interpreting observables, so that Question and Answer 5.18 are the core of this section. However, cleavage is so common a feature of deformed rocks that refraction of cleavage deserves mention, as follows.

Refraction of Cleavage   Cleavage is the tendency of a rock to split open most easily along one particular set of close-spaced parallel planes. Examples of textures that give rise to cleavage are shown in idealized form in Figure 5.14. If a rock undergoes coaxial deformation (one line of particles continuing all through the deformation to be the line of maximum strain rate) one can expect a simple relation between the cleavage direction and the stress field; usually, the cleavage plane is normal to the direction of greatest compressive stress and greatest linear shortening. But as soon as one comes to noncoaxial deformations, all simplicity disappears: in some circumstances, a cleavage can behave roughly like a passive marker, but in other circumstances it can behave more like e_{max} or $\dot{e}_{max}$, and statements about one of the cleavage types in Figure 5.14 are not likely to apply to the other types. Some outcrops show cleavage and other features that make it reasonable to put forward a conjecture about how the cleavage attained its present

a   b    c

FIGURE 5.14

orientation; but to generalize from these instances is not warranted. The single term "cleavage" makes it *seem* as if we are dealing with a single simple phenomenon, but we are not: to think of two cleavages as having natures and histories as different as if they were two people is probably more realistic. One can *imagine* simple behaviors and relations, but alas, cleavages, like people, tend not to stick to them. (The topic is continued on page 138, under the heading Behavior of Multilayers.)

ALTERNATING MULTILAYERS AND ANISOTROPY

Of rocks in the top 20 km of the earth, many have been sediments at one time or another, and sediments are by nature multilayered. On a smaller scale, rocks also develop cleavage, and some cleavages are in fact due to alternating compositional layers, as in Figure 5.14(c). Hence we need to consider how layering affects a rock's mechanical behavior.

Natural layering in sediments can be quite complicated and nonrepetitive, and hence difficult to treat by any simple theory. But an idealized alternation of layers can be treated: without being too complicated, such alternating multilayered sets bring to light ways in which layered materials differ from unlayered ones. Multilayer behavior has already been touched on at several points, and we begin by bringing together results from Answers 5.4 and 5.5 (shearing) and 5.13 (flattening), as follows.

The assembly discussed consists of alternating layers of shale (thickness 12 cm, viscosity 30 gvu) and limestone (thickness 8 cm, viscosity 80 gvu). When subjected to a shear stress parallel to the layering, the effective viscosity of the assembly is 40 gvu (see Question 5.5), but when subjected to compressive stresses normal and parallel to the layering, the effective viscosity has a different value.

Question 5.22 Weighted mean viscosity: What is the effective viscosity of the assembly in the conditions of Question 5.13? To use the formula viscosity = (stress difference)/(4 × linear strain rate), we need a single effective or average stress in place of the separate stresses 40 MPa and 6 2/3 MPa in the shale and limestone layers, so proceed by first calculating an effective or average stress. (We need a *weighted* mean as in Equation (3.3).)

The conclusion from Question 5.22, that the assembly behaves more stiffly if we try to flatten it than if we try to shear it parallel to the layering, is perfectly general: every layered succession has this property. The difference can be more striking, and it affects real processes such as the development of folds; thus we consider it in a little more detail, as follows.

First, there are short cuts from the separate layer viscosities to the effective viscosities of the assembly. The **weighted mean viscosity** is

$$(3/5) \cdot (30 \text{ gvu}) + (2/5) \cdot (80 \text{ gvu}) = 50 \text{ gvu} = \text{effective viscosity for flattening.}$$

Also, if we use the idea of fluidity as the reciprocal of viscosity, the **weighted mean fluidity** is

$$\underbrace{\left(\frac{3}{5}\right) \cdot \left(\frac{1}{30}\right)}_{\substack{\text{fluidity of} \\ \text{shale}}} + \underbrace{\left(\frac{2}{5}\right) \cdot \left(\frac{1}{80}\right)}_{\substack{\text{fluidity of} \\ \text{limestone}}} = \frac{1}{40} = \underbrace{\frac{1}{\text{effective viscosity}}}_{\text{for shearing}}.$$

That is to say, in layer-parallel flattening, the effective viscosity is the weighted mean viscosity, while in layer-parallel shearing, the effective fluidity is the weighted mean fluidity.

Question 5.23 Weighted mean viscosity, continued: Verify these relations, either by algebra or by inventing a completely fresh numerical example and checking out the assertions just made.

Second, we recall and emphasize that shearing involves linear strain and linear strain involves shearing. The shear stress of 24 MPa parallel to the layering in Figure 5.13 cannot exist in isolation: it can only exist as part of a complete stress state, as represented by a sine-wave diagram or the dashed-line circle in Figure 5.15(a). To avoid tangling with the normal stresses of 60 MPa, 40 MPa, and so on, the simplest thing is to imagine the shear stress as part of the stress system in Figure 5.15(b), which is represented by the solid-line circle in diagram (a). This gives a slightly different view of the behavior just noticed: for the alternating limestone and shale assembly, if we put in principal stresses at 45° to the layering, we find an effective viscosity of 40 gvu, whereas if we put in

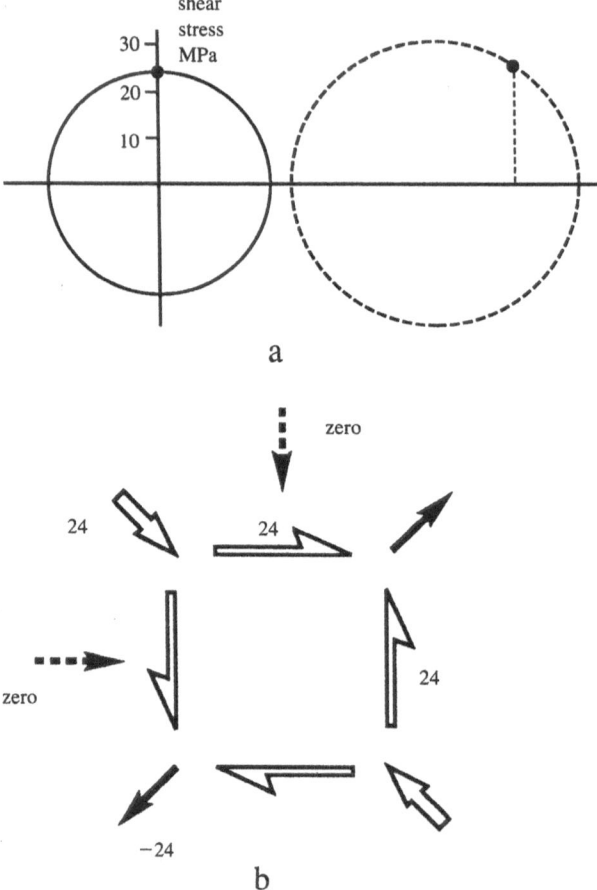

FIGURE 5.15

principal stresses at 0° and 90° to the layering, we find an effective viscosity of 50 gvu. We do not put in *either* shear stresses *or* normal stresses—it is impossible to put in one without the other. It is the *orientation* of the stress system that properly distinguishes the two cases. So we come to the concept of **anisotropy**: the same sample of material shows different properties according to which direction in the material is subjected to an outside influence—in our case, compression. Materials can also be anisotropic for transmission of light, thermal conductivity, magnetic state, and so on.

Question 5.24 Weighted mean viscosities, continued: Suppose that in Figure 5.13, the material above the interface was not pure shale but was a fine interlamination of shale and limestone layers in proportions three-fifths shale and two-fifths limestone. What linear and shear strain rates would be observed parallel to the layering?

The simplicity of Answer 5.24 is noted. It derives from concentrating attention on directions parallel and perpendicular to the layering in the material. Geological situations often involve layering that is inclined to the horizontal and principal stresses that are parallel to the horizontal (see for example Figures 4.3 and 5.6). It is not efficient to try to work out how the layered material responds to a principal stress at some oblique angle to the layering; it is more efficient to break up the stress state into components parallel and normal to the layering. We switch to what may seem a more complicated description of the stress state, for the sake of using a simple picture of the material's response.

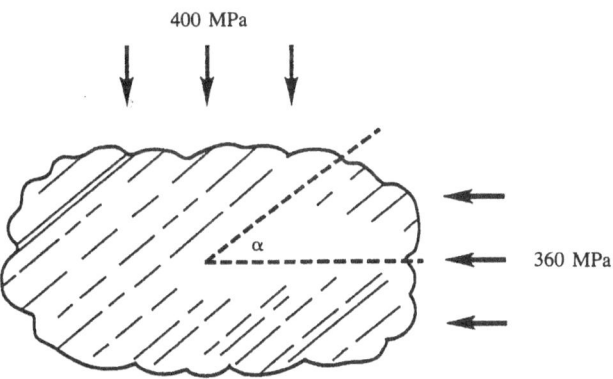

400 MPa

α

360 MPa

FIGURE 5.16

Question 5.25 Behavior of an anisotropic material: A mica schist has viscosity 120 gvu for linear strains parallel to the schistosity and 30 gvu for linear strains at 45° to the schistosity. What is its response in the situation in Figure 5.16? Take sin α = 0.6 and cos α = 0.8.

The conclusion in Answer 5.25, that the shear strain rate is almost 30 times the linear strain rate, could be geologically realistic, as far as anyone knows. The difference results partly from the schistosity or anisotropy, the factor of 4 difference between 120 gvu and 30 gvu inherent in the material, and partly from the *orientation*. If the schistosity and the stress field were differently oriented with respect to each other, the ratio of strain rates could be smaller, or larger still.

Summary

1. If a sample of material observed on one scale is layered or schistose when observed on a smaller scale, its response to stress will vary according to the relation of the stress directions to the texture (layering or schistosity).

2. The effective viscosity of a composite material is greatest when the principal stresses are parallel and normal to the layering or schistosity, and least when they are at 45° to the same.

3. The ratio of greatest to least effective viscosity is a measure of the extent to which the material is mechanically anisotropic. Two materials can be equally anisotropic even if one is, measure for measure, 10 times more viscous than the other.

4. If, in an alternating layered assembly, a fraction h has viscosity M gvu and a fraction k has viscosity N gvu, with $h + k = 1$, the greatest and least viscosities of the assembly are $hM + kN$ and $1/[(h/M) + (k/N)]$ gvu.

5. The response of an anisotropic material to stress is best understood by expressing the stress state in terms of two normal stresses (parallel and perpendicular to the layering or schistosity) and a shear stress; the material responds to the normal stresses with its maximum viscosity and to the shear stress with its minimum viscosity.

6. The principal directions of the resulting strain-rate condition are not parallel to the principal directions of the driving stress field, except when the latter make angles of 0°, 45°, or 90° with the layering or schistosity.

Angular Relations   The emphasis in the preceding description has been on angles of 0°, 45°, and 90°; other angles lead to the calculations becoming more complicated. But the last statement in the Summary deals with other angles and is given without any back-up; it is interesting to explore the extent to which it is true.

Question 5.26 Principal directions of strain rate: What are the principal directions of the strain-rate response in the layered material from Question 5.24? (The directions of $\dot{e}_{max}$ and $\dot{e}_{min}$ can be found by a construction exactly like that used for σ_{max} and σ_{min} in Answer 4.10.)

Question 5.27 Principal directions of strain rate, continued: What are the principal directions of the strain-rate response for the schist in Figure 5.16? Use rates from Answer 5.25; mark the direction of greatest elongation rate on the figure.

The last two examples give angular separations of less than 10° between principal directions of strain rate and stress, and probably in geological materials the separation does not often become greater than this. A nongeological extreme case is shown in Figure 5.17: a tall stack of glass microscope slides is tilted to, say, 3° and compressed from the sides. The only strain the assembly can undergo is slip of one slide on another, giving $\dot{e}_{max}$ for the assembly at 45° to the slides as discussed in Chapter 2 (Figures 2.29 and 2.31). The least compression, σ_{min}, is vertical or at 87° to the slides, and thus is at 42° to the direction of $\dot{e}_{max}$. Geological materials are probably never anywhere close to this high degree of anisotropy, though it is a convenient extreme case to carry in the imagination, as will be seen in the section that follows.

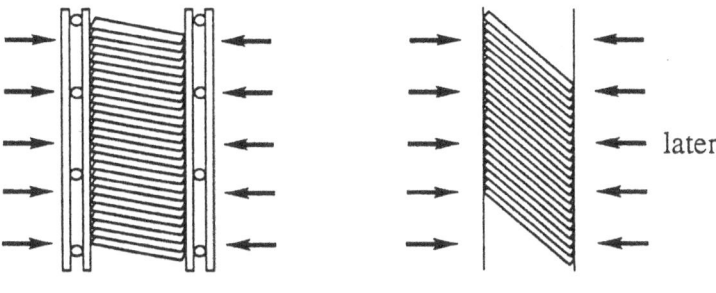

FIGURE 5.17

STRAIGHT-LIMBED FOLDS

Early investigators of geology faced two rather prominent questions about rock deformation. If one finds a seashell embedded in the rock on top of a mountain, one may wonder how it got there; and if one sees ordinary layered rocks thrown into outcrop-sized folds, one may wonder how they became crumpled. Some observers have been struck with the thought, "What *enormous forces* must have been at work" but another response, perhaps even closer to the heart of the matter, is, "What a *long time* that must have taken." An example of fold development is shown in Figure 5.6. If the fold had had the more open profile at the time mankind first learned to control fire, it could have closed to the more acute profile by today. A fold may not go through its rather simple geometrical evolution any faster than the average human skull has changed proportions between Paleolithic times and now.

Besides long times and large forces, another mind-stretching attribute of folds is their variety, and even, in some outcrops, their complexity. But to piece together an understanding, the way to begin is with a simple example, as in Figure 5.18.

The advantage of the situation shown is that we know most of its behavior already. Assume that the material in view is the schist from Question 5.25. Compared with that question, the stress magnitudes of 360 and 400 MPa have been interchanged, but this simply reverses the strain rates: we now expect line p to *shorten* at 0.023 per gtu, and q to *elongate* at the same rate; the pair pq still suffers shear strain at 0.64 per gtu. But

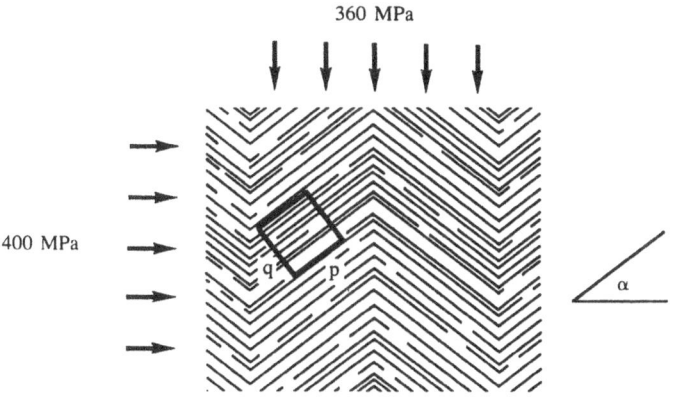

FIGURE 5.18

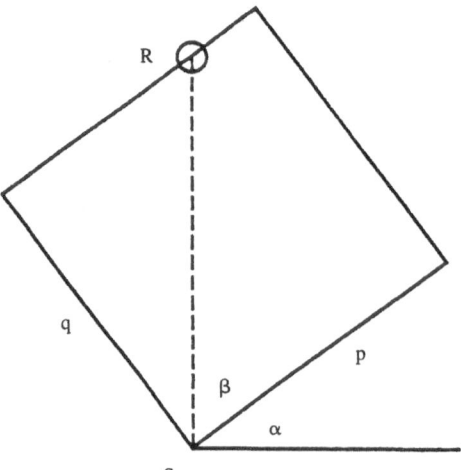

Figure 5.19

surely also, the folds will close; the kinking will become more acute; the angle α will increase. How is the *rate of increase of* α related to the strain rates we already know?

To answer the question, consider a square portion of schist as in Figure 5.19. Each of the strains—shortening of p, elongation of q, and change of the angle pq—will displace the point R. We are not interested in R's vertical motion, but any horizontal motion of R will change the angle β. In Figure 5.19 it is convenient to think of line p and angle α as staying fixed, and line SR rotating about S, though if we go back to the overall setting, Figure 5.18, it is of course RS that stays still (while elongating vertically) and line p that rotates.

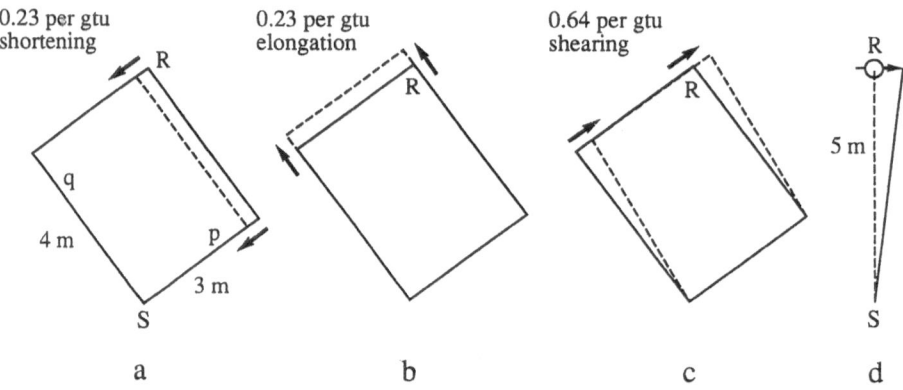

Figure 5.20

Question 5.28 Rotation rate during folding:

(a) What is the displacement rate or velocity of R in each of the separate processes a, b and c in Figure 5.20, in meters per gtu?

(b) What is the horizontal component of R's velocity in each process?

(c) What is the net horizontal velocity of R?

(d) What is the rate of rotation of RS about S, in radians per gtu?

The exercise just completed is the starting point for several trains of thought. First, if we came back 1 gtu later, would we find the schistosity inclined 23° steeper? No, we would not: the calculation is very much tied to the *current* value of angle α, and as α changes, every strain rate and angular ratio changes. To calculate behavior at $\alpha = 8°$ for example, we would have to repeat Answer 5.25 to get new strain rates, and then redraw Figure 5.20 to show a rectangle with a much smaller width/height ratio. The following approximations serve: $\sin \alpha = 0.14$, $\cos \alpha = 0.99$, $\sin 2\alpha = 0.28$, $\cos 2\alpha = 0.96$, and rectangle sides could measure 1 m along p and 7 m along q. Linear strain rates are larger, 0.08 per gtu instead of 0.023; the shear strain rate is smaller, 0.19 per gtu instead of 0.64; and the rotation rate or rate of change of α is smaller, 10° per gtu instead of 23°. (*Check:* What is the linear strain rate when $\alpha = 0°$? *Answer:* (40 MPa)/4 $\times$ (120 gvu) = 0.083 per gtu, so a *slightly* smaller number at $\alpha = 8°$ checks out. Ideally, the shear strain rate and rotation rate would be zero at $\alpha = 0°$, but a deflection of α of just a few degrees is enough to get the shearing and rotation processes started.)

The purpose of this text is practice mechanics and not to explore the history of folds, but the reader can get results for $\alpha = 45°$, 53°, 60°, 82°, and so on if desired; it is interesting to see how the rate of rotation climbs to a maximum and then diminishes again. Going back to Figure 5.18, we can also consider the overall horizontal and vertical dimensions: their rate of change also climbs to a maximum value and then declines.

Question 5.29 Elongation rate in folding:

(a) Compared with Question 5.28 part (b), what is the *vertical* component of R's velocity in each process, and the total vertical velocity of R?

(b) What is the linear strain rate of *RS?* What is the effective viscosity of the whole assembly in Figure 5.18?

Rounded Folds and Saddle Reefs   It has just been noticed that if schistosity is parallel to the maximum compression (horizontal in Figure 5.18), the material behaves with viscosity 120 gvu, whereas if schistosity is at 37° to the maximum compression, the effective viscosity for horizontal shortening is only 32 gvu. This information permits us to consider the situation in Figure 5.21.

Question 5.30 Stresses in the hinge of a fold: What stress *S* MPa permits the center section to elongate vertically at the same rate, 0.31 per gtu, as the limb sections are elongating?

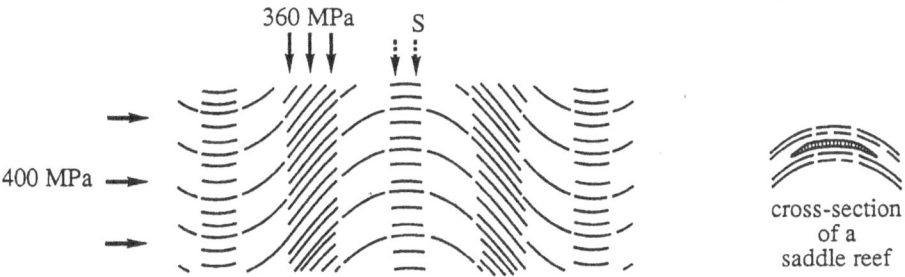

cross-section
of a
saddle reef

FIGURE 5.21

It *appears* that we have a simple and complete analysis of what vertical stresses such as S are needed at any point along the top edge of Figure 5.21; even at a point in the "shoulder" of the fold, where the schistosity is inclined at some angle between zero and 37°, the methods already used would permit calculation of a stress S' that would allow vertical elongation at 0.31 per gtu. But geological reality is not so simple. It is important to remember that the relation

$$\text{linear strain rate} = (\text{stress difference})/4 \times \text{viscosity}$$

applies only when the area of rock in the plane observed stays constant, i.e., when $\dot{e}_{max} = -\dot{e}_{min}$. In reality, if a layer of rock subject to 360 MPa of compression is next to a layer subject to compression that is 109 MPa less, it is *highly* likely that *some* material will find a way to migrate from the high-compression zones into the zones of lower compression. A similar point was made on page 120 in connection with boudinage (in fact, the center section in Figure 5.21 is effectively stiffer than the sections on either side just as if it were a vertical stiff rock layer flanked by layers of a less stiff rock type). Material migration is discussed quantitatively in the next section; here three points are worth noting:

1. The effect is that the vertical compressive stress in a real schist would be less than 360 MPa in the limb regions of a fold, and more than 251 MPa in the horizontal crest region; the effect of material migration is to diminish the stress contrast between limbs and crest.
2. Migration can be effective over a few millimeters, but not so commonly over kilometers: the ratio (width of crest region)/(width of limb region) can be around 1 where the dimensions are in millimeters, but where the limbs are longer, the ratio is smaller; i.e., the crest regions are *proportionately* narrower, and look sharper, as in Figure 4.3.
3. The material migration has economic consequences. It may lead to crescent-shaped pockets of new material in the crest regions, often mainly quartz; these sometimes contain gold or other materials of value. Being three-dimensional, the pockets are more like saddles than crescents and are known as saddle reefs.

Behavior of Multilayers

An essential point from the preceding section is that a thick stack of alternating layers behaves like a schist: its effective viscosity for linear strain parallel to the layering is greater than its effective viscosity for linear strain at 45° to the layering (i.e., for shear strain parallel to the layering); in the shale–limestone example used, the maximum and minimum viscosities were 50 and 40 gvu. To describe the *overall* behavior of such an assembly, those two viscosities are the only information we need; they permit calculation of the manner in which a straight-limbed fold would close by exactly the method in Answer 5.28. There is an internal detail of interest, however: the overall geometry forces the shale and limestone layers to *shorten* at the same rate (linear strain parallel to the layering), but it leaves them free to *shear* at different rates (Figure 5.22). At least, at a sufficient distance down either limb away from a crest, the shale and limestone can shear differently; in the crest, interactions are more complicated but for a typical straight-limbed fold, there is a large section between crest and trough where behavior as in Figure 5.22(b) can develop as if the limb were infinitely long.

For the behavior in Figure 5.22(b) to be observable, of course, a marker is needed. Too many shale and limestone layers lack the needed markers, but worm tubes or dessication cracks are present in a few folded assemblies. These naturally tend to be at

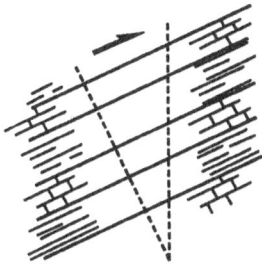

a   averaged behavior

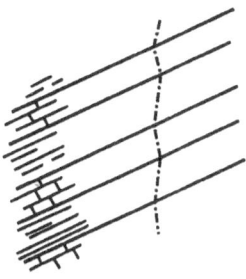

b   actual behavior

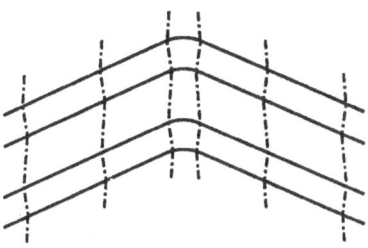

c   in the crest, alternate
     layers show fanning
     and anti-fanning

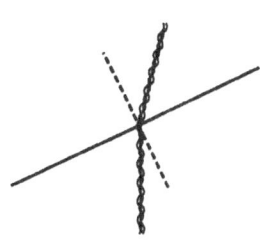

d

FIGURE 5.22

about 90° to layering in undisturbed sediments. If the sediments were then folded in the accordion manner shown in Figure 5.18, the refraction of the markers would reflect the relative viscosities of the layers, as discussed in Question and Answer 5.19.

Question 5.31 Estimating viscosity ratio from observables:

(a) What is the apparent viscosity ratio of the two rock types in Figure 5.22(d)?
(b) What errors or interfering effects is the estimate in part (a) subject to?

(c) If one of the rock types were a shale or schist, it would have a maximum and a minimum viscosity, as discussed in the previous section. What does the ratio estimated in part (a) signify in this circumstance?

Aside from mud cracks and worm tubes, there is a possibility that early deformation may create features that serve as markers in later deformation. Figure 5.23 suggests that in the early stages of fold development, when layer-parallel shortening is at its maximum intensity, limestone layers may develop solution seams and shale layers may develop microfolds. In later stages, these relics of early layer-parallel shortening cease to develop, and serve simply as passive markers of later layer-parallel shearing. They are not ideal markers however, because even though they develop *early* they are not likely to develop quite at 90° to the layering.

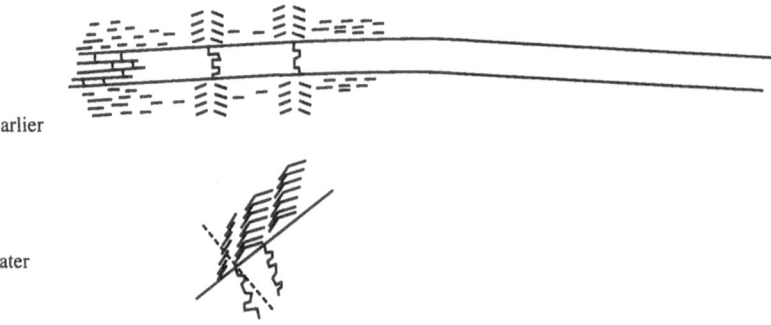

earlier

later

FIGURE 5.23

The preceding remarks show that a set of alternating layers has potential for a more complete mechanical analysis than a schist. No amount of inspection of a schist will reveal the ratio of its maximum to its minimum viscosity; but in a multilayer with markers, we can measure shear angles to estimate a viscosity ratio and also measure thickness fractions (2/5 and 3/5 as discussed above, or more generally h and k as in Answer 5.23). From these, the formulas

$$\text{maximum viscosity} = hM + kN \tag{5.4}$$

and

$$\text{minimum viscosity} = 1/[(h/M) + (k/N)] \tag{5.5}$$

give the effective properties of the assembly, at least relatively—if we assume that one rock type has viscosity N gvu, all other viscosities can be expressed as multiples of N. The ideal set of alternating marker-rich folded layers that will allow the theory to be checked in all details has not yet been found. Nonetheless, if we go back to Figure 5.4 and ask, Where in the real world can we find an example of the operation of this small, compact set of concepts? an example of straight-limbed folding can be offered as an answer.

CHANGE OF VOLUME

Early in this chapter, where the ability of normal stresses to produce linear strain rates was first noticed, we made the assumption that the volume of the sample considered did not change. As the subsequent sections have shown, many questions about the earth's behavior can be explored using that assumption. On the other hand, under some circumstances there is no question that the volume of a body of rock does change; the purpose of this section is to consider what type of quantitative statements can then be made. Two examples are wet mud that undergoes compaction, and an almost dry granular rock whose grains diminish as thin films of fluid migrate along grain boundaries carrying material away in solution. At first sight, it seems likely that the latter process would be slow and ineffective. It *is* slow but is not necessarily ineffective—it seems not uncommon for half a rock's initial volume to be carried away in such a manner.

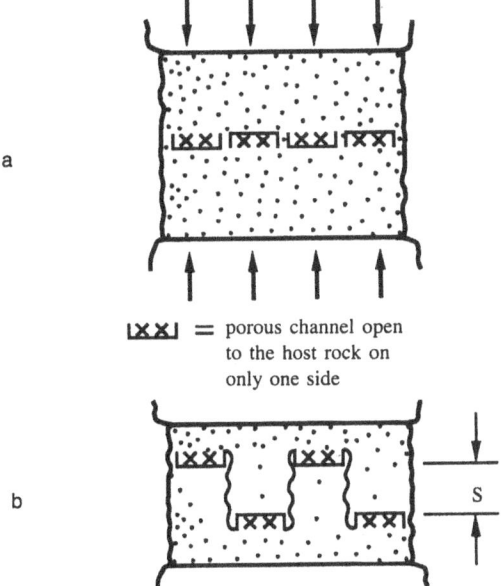

FIGURE 5.24

Question 5.32 Stylolites: Imagine a soluble rock heavily compressed from above and pierced by porous channels through which a solvent is pumped (Figure 5.24). Each channel is open on only one side, either top or bottom, so that the dissolving action is similarly one-sided. After some while, the geometry of the sample might be as in Figure 5.24(b), and an observer could use the distance S as an estimate of the thickness of rock that had been lost in solution.

(a) Apply the train of thought just introduced to Figure 5.25. Estimate the ratio (thickness of rock lost)/(thickness now remaining).

(b) Public buildings often incorporate pieces of stylolitic limestone. Repeat part (a) for some piece of real rock. Suppose the visible surface is not cut perpendicular to the dark seams where dissolution occurs: does this cause error in your estimate of the ratio?

10 mm

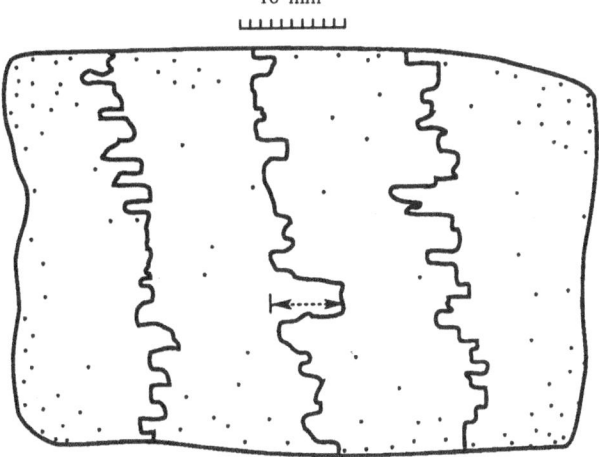

FIGURE 5.25

Note that, assuming the other two dimensions of the sample are not affected by the solution process,

$$\frac{\Delta(\text{thickness})}{(\text{thickness})} = \frac{\Delta(\text{area})}{\text{area}} = \frac{\Delta(\text{volume})}{\text{volume}};$$

$$\text{linear strain} \quad = \quad \frac{\text{area}}{\text{strain}} \quad = \text{volume strain}.$$

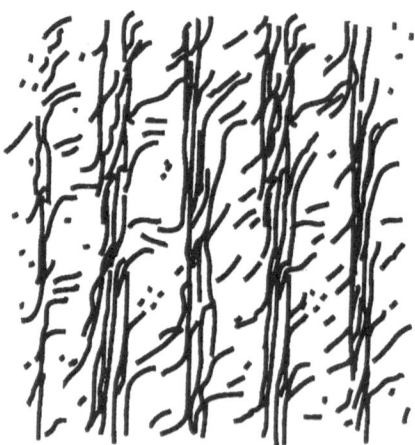

FIGURE 5.26

Question 5.33 Insoluble residues: Figure 5.26 shows a rock composed of laminae for which the following are average values:

mica-poor laminae	thickness 3.2 mm mica content 12 percent, 0.12;
mica-rich laminae	thickness 1.8 mm mica content 72 percent, 0.72.

If the rock was initially homogeneous, with the composition of the present mica-poor laminae, and the mica-rich laminae developed by *removal* of non-mica constituents in solution, estimate the volume lost, as a fraction of the volume originally present.

The laminae in Question 5.33 are coarser than average, but evidence increasingly suggests that most slates and many shales have been subjected to at least some selective dissolution of the same type on a finer scale.

In both the preceding exercises, change of volume is *estimated* using some assumption about the process that has occurred; we have no direct knowledge of the sample's original dimensions. Circumstances where the original dimensions are known are in fact exceedingly rare: the best-known instance involves graptolites. These small soft animals seem to have the property that a species has a "proper length" for the spacing of its cups (see Figure 5.27(a)); every member of the species builds cups at this spacing whether young or old, healthy or sick, regardless of environment—a tapemeasure animal.

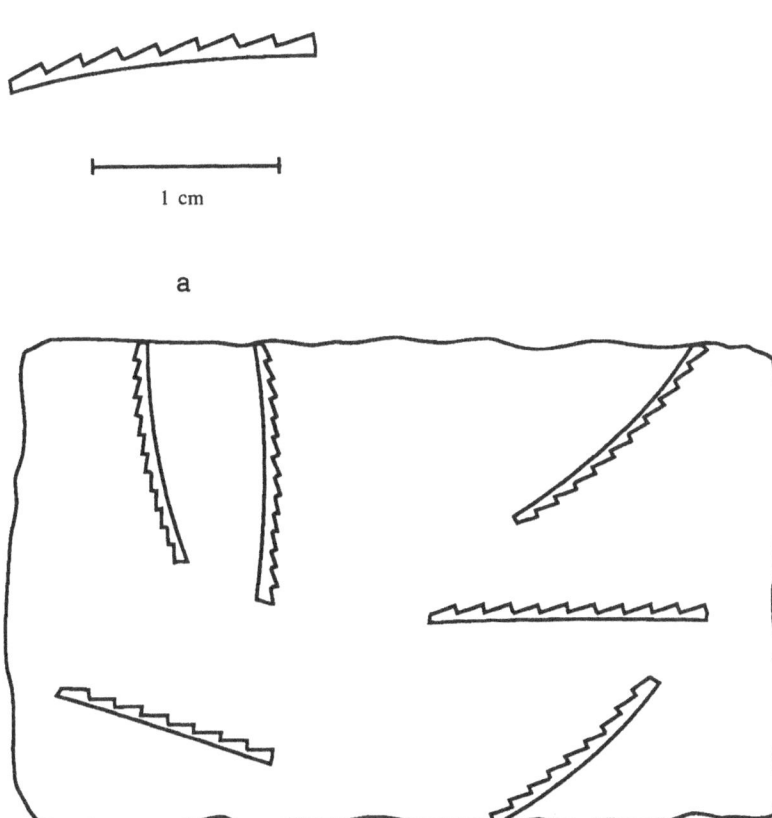

Figure 5.27

Question 5.34 Graptolites: The graptolite in Figure 5.27(a) is undeformed. If the graptolites in diagram (b) are the same species, how does the *area* in diagram (b) compare with its original area? (*Suggestion:* Count the number of cups per centimeter on each column.)

The three preceding exercises are based on real examples, and change of volume by factors between 0.5 and 1.5 cannot be considered uncommon. Hence all the equations and estimates in earlier parts of this chapter need to be reconsidered. Conservation of local areas and volumes is all right as an *initial* assumption, but an early next step should be to question the validity of the assumption and, if possible, estimate the amount of volume change.

Permeability

As just introduced, the idea of change of volume is linked to the idea that part of the sample that was initially present "escapes." At any moment, the material in view has two parts (see Figure 5.28) and one part is escaping by traveling through the other part. Figure 5.28(a) is the simpler: the two parts are "grains" and "fluid"; the fluid escapes

a

insoluble
grains

b

soluble
grains

earlier:

later:

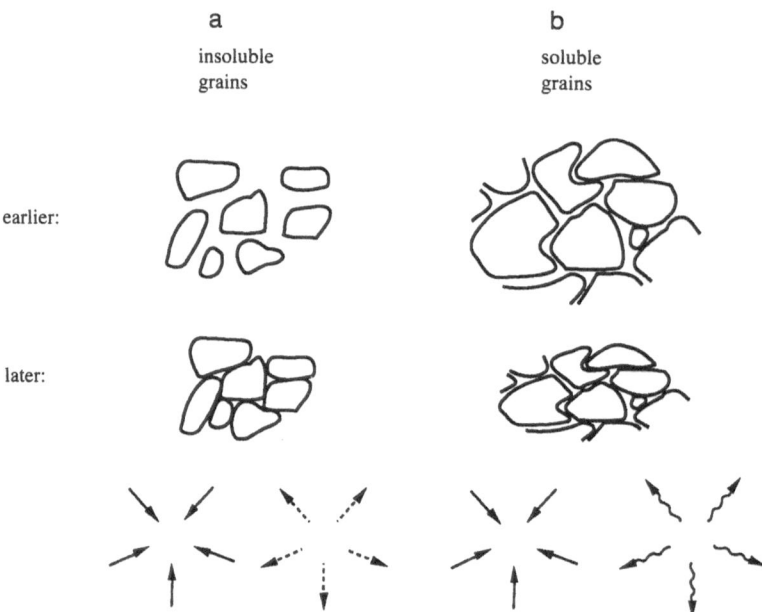

FIGURE 5.28

and the grains retain their original shapes while clustering more closely together. Figure 5.28(b) taxes the imagination slightly more: the intergranular channels are shown as maintaining their width, and it is the *grains* that occupy less space at the later stage. The part that remains is "grains" but the part that has escaped is also primarily grain material—quartz or calcite or whatever it might be. Even so, at any moment, there is a clear separation into two parts—the quartz that is in process of staying within the sample considered and the quartz that is in process of escaping from it. The difference is that in Figure 5.28(a) "once a grain, always a grain"—each particle of material has a role, either to stay or to escape, and no particle changes its role; whereas in Figure 5.28(b), a speck of quartz that *stays* during one time interval may become a migrant in a later interval. We should not let this fact confuse us; it is as true for (b) as it is for (a) that, at any moment, the sample is clearly divisible into a part that is staying and a part that is escaping. As noted before (page 123), the cause-and-effect relations we are pursuing are relations between the cause present at some moment (the stresses) and the effects resulting at that moment (particle displacement rates), without any connection to earlier or later times; hence the equations that we need in order to get started are very much the same for situation (b) as they are for situation (a).

It is to be admitted that an *extra* process occurs in (b), namely the conversion of some material particles from being "stayers" to being "escapers"—i.e., the process of dissolution. But this process is driven mainly by chemical influences such as undersaturation; the physics, that drives the escapers along channels among the stayers, is much the same for (b) as for (a).

Flow Rates   The preceding paragraphs established that a geological sample can often be divided into "stayers" and "escapers"—sometimes, but not always, solid and liquid, respectively. If we have a fixed reference frame, we can define some plane and ask, "At what rate is solid matter crossing the plane?" and "At what rate is fluid crossing the plane?" For example, for 1 km^2 of horizontal plane inside a pile of sediments, 5 m^3 of solids might cross downward and 5 m^3 of water might cross upward in a year, the plane being defined with respect to sea level, or the center of the earth, or some such external standard.

Related statements would be: "the solid is flowing downward at 5 μm per year" and "Water is flowing upward at 5 μm per year." These are fictional flow rates that would give 5 m^3 of material if they occurred uniformly over the whole 1 km^2. More real flow rates depend on the relative volumes: if the pile were 80 percent solid grains and 20 percent water, we could imagine grains traveling down at 6 1/4 μm per year across 4/5 km^2 and water traveling up to 25 μm per year across 1/5 km^2.

It is also useful to consider how fast water and solids are traveling with respect to each other; the relative velocity is 31 1/4 μm per year using the "real values" and 10 μm per year using the fictional values. The 31 1/4 is not quite true because the 25 and 6 1/4 are only average values, subject to much actual local variation. The 10 μm is quite untrue and is based on the totally impossible concept of water occupying the whole space and sweeping upward, while simultaneously rock particles also occupy the whole space and sweep downward. In spite of this quality, the 10 μm per year is a useful number for several practical purposes: it is called the Darcy velocity.

> The **Darcy velocity** is the velocity of fluid and matrix *relative* to each other, expressed *as if* the fluid and the matrix both simultaneously occupied the whole of their joint space.

Permeability and Darcy's Law   Darcy's law states that if there is a pressure gradient in the fluid of H (Pa/m or MPa/km, etc.), the Darcy velocity is proportional to H. The proportionality constant is the system's **transmissivity,** here designated K. Thus

$$\underset{\text{m/sec}}{\text{Darcy velocity } V} = \underset{\text{m}^2/\text{Pa-sec}}{K} \cdot \underset{\text{Pa/m}}{H}. \tag{5.6}$$

It is helpful to regard this equation as a definition of the transmissivity K at any instant, regardless of its value at another instant; then Darcy's Law states that K does not change when H changes. Thus we separate the definition (always exactly true) from the empirical law (almost exactly true for some materials, but far from true for others, and never wholly true).

Note three related quantities:

$$\text{transmissivity } \frac{\text{m}^2}{\text{Pa-sec}} = \frac{\text{permeability (m}^2)}{\text{viscosity (Pa-sec)}} \quad \begin{array}{l}\text{of matrix}\\ \text{of moving fluid}\end{array}$$

$$= \frac{\text{hydraulic conductivity (m/sec)}}{\text{fluid density} \times g \text{ (Pa/m)}} \tag{5.7}$$

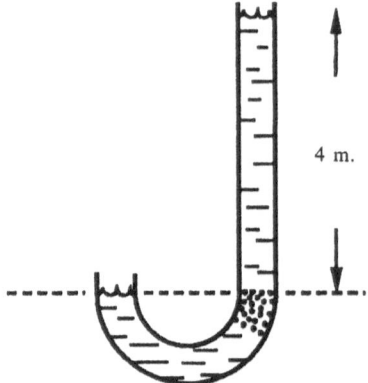

4 m.

FIGURE 5.29

Question 5.35 Flow through porous material: If a pipe as in Figure 5.29 is clogged with a plug 5 cm long whose transmissivity is 12.10^{-8} m²/Pa-sec, how fast will water flow upward at the lower end of the tube? Could you move your finger at the same speed?

This is the *fictional* flow rate through the plug *as if* water occupied the full width of the pipe where the plug is in place. It is also the real flow rate in the lower bend, because in the bend the water does occupy the full width of the pipe. Given the transmissivity, we never need to inquire what the actual pore space is in the plug, nor the real velocity through the pores; the transmissivity and the Darcy velocity summarize the *practical effects* of the plug, and successfully bypass the difficult task of describing the plug's detailed interior geometry.

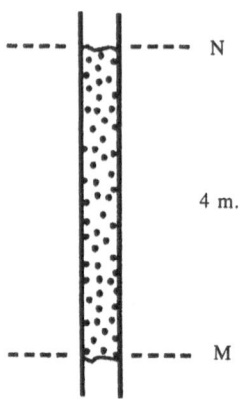

N

4 m.

M

FIGURE 5.30

Question 5.36 Pressure gradient driving flow: See Figure 5.30. If a full 4 m of pipe is plugged with material of transmissivity 12.10^{-8} m²/Pa-sec and the pressure at M is 0.04 MPa greater than at N, at what rate will water flow up the pipe? (*Warning:* This is a trick question.)

Question 5.37 Flow through porous material, continued:

(a) If the 4 m of plugged pipe were laid *horizontally* and pressure at *M* were 0.04 MPa greater than at *N*, at what rate would water flow *along* the pipe?

(b) If the plugged pipe were returned to vertical and the pressure at *M* was equal to the pressure at *N*, at what rate would water flow down the pipe (e.g., if the pipe had open air at top and bottom, and water flowed down simply because of its own weight)?

(c) What pressure difference between *M* and *N* would make water flow up the pipe at the rate you calculated at (a) and (b)? (Parts (a) and (b) have the same answer: do you agree?)

A conclusion is that the answer given to Question 5.35 is not quite exact. The pressure gradient in the plug is actually 0.79 MPa/m (decreasing downward) but a hydrostatic gradient is 0.01 MPa/m (decreasing upward) so the flow-driving gradient (the difference between these two) is 0.80 MPa/m as stated.

Nonuniform Pressure Gradients

Darcy's law relates the local flow rate across a particular imaginary plane to the local pressure gradient at the same plane. If we think of two planes 1 m apart, whether in loose granular material or in "solid rock," there is the possibility that the flow rate across plane 1 is not equal to the flow rate across plane 2. The purpose of this section is to give a quantitative description of such imbalances and to consider their geological consequences.

Column of Sediments   For the sake of a real-life example, we consider a location such as the Dead Sea rift, the Bay of Bengal, or the Mississippi delta, where sediments have accumulated to a thickness of say 6 km or 20,000 ft. Question 3.5 begins the task of estimating pressures in such a pile: if there were an unobstructed water-filled hole reaching to the bottom, the pressure profile in the hole would match the hydrostatic line in Answer 3.5. On the other hand, if water were *trapped* in intergranular pores and wholly unable to escape toward the surface, its pressure would be close to the pressure of the overburden (i.e., water pressure = lithostatic pressure). Even unconsolidated sediments have a bulk density of more than 2 g/cm^3 or 2 t/m^3 so the profile of water pressure with depth would be more than twice as steep (see Figure 5.31).

In reality, water near the top of such a pile has fairly ready access to the surface (i.e., conditions are likely to approach hydrostatic), whereas water at the base of such a pile is far closer to the condition of being trapped, and so in very general terms we expect a profile of the type shown. There is likely to be a middle section with particularly rapid change of pressure with depth, the profile being concave to the right above the inflexion and concave to the left below it.

Suppose first that the permeability of the sediment-pile is the same at all levels, and that the pile is broad and flat so that escape of water is upward, uncomplicated by migration sideways. Then level *G* would be drying out; the gradient driving water to level *G* from below is less (in MPa/km) than the gradient driving water away from level *G* to higher up. This is reasonable since buried sediments do undergo compaction. Indeed one might argue that, because we know compaction occurs, we should expect at *every* level to find more water flowing away upward than flows in from below. This would require the whole curve to be composed of many corners concave to the left,

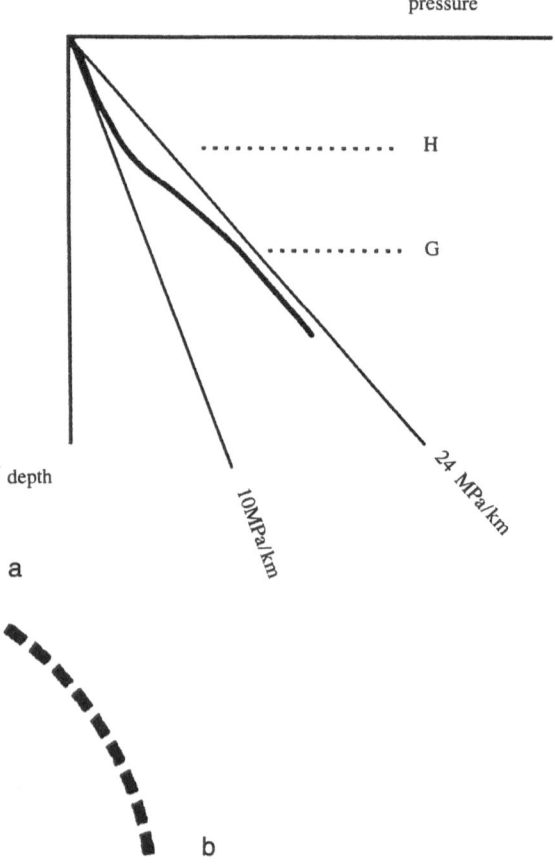

pressure

H

G

depth

10MPa/km

24 MPa/km

a

b

FIGURE 5.31

Figure 5.31(b), or a smooth concave curve, as the *low part* of the main profile seems to be. But what geological processes occur where the curve is concave to the right?

If the curve truly has this shape, then if the permeability were uniform the sediments at level *H* would necessarily be filling up with water; any time a flow carries fluid into a region of diminished flow rate, the flow "backs up", i.e., fluid accumulates. Actually the permeability is likely to be far from uniform: all the way up a pile of uniform sediment type, the permeability tends to increase, and at every level, the product (permeability) × (pressure gradient) is probably increasing upward. Thus, at every level, the flow out upward is greater than the flow in from below, and compaction goes on. But in nonuniform sediments, the opposite can occur—and decompaction (an *increase* in the pore space between grains) is almost certainly a real process affecting parts of a real pile, probably in spasms rather than in a steady, continuous way.

The purpose here is not so much to reconstruct the fluid-pressure history of compacting piles (though the topic is of great interest and enormous economic consequence in the oil business); the purpose is rather to emphasize the role of **change of flux.** A uniform flow of *X* through *Y* leaves *Y* unaffected: it is *imbalance,* a difference between the amount flowing in and the amount flowing out, that has practical consequences for *Y*.

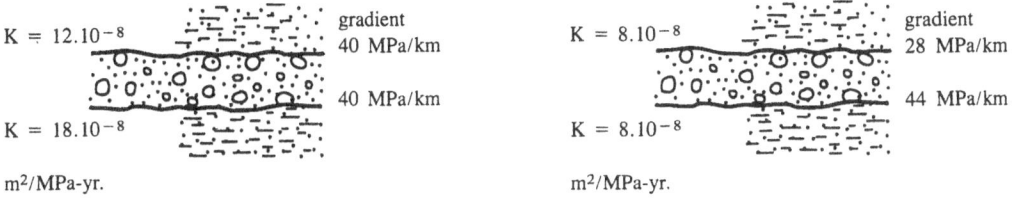

FIGURE 5.32

Question 5.38 Unequal flows in and out: Both of the layers marked with open circles in Figure 5.32 are increasing in fluid content (more fluid flowing in than flowing out). (The gradients shown are the driving gradients; the hydrostatic gradient has already been subtracted.)

(a) Which is increasing at the faster rate?
(b) Convert the faster rate into centimeters per million years (cm/my), and into a strain rate per gtu, taking the layer thickness as 4 cm.

Generalization   There are parallels between Question 5.38 and flow of heat. Moreover, we shall need the same train of thought at a later time without the sedimentological details, and hence it is worth restating the exercise in general symbols.

Let us designate the base of the layer by suffix B and the top by T. Then

$$\text{flow in} = K_B \cdot \frac{(dP)}{(dh)_B} \text{ m/yr} \quad \text{or} \quad \text{m}^3/\text{m}^2\text{-yr}$$
$$\underset{\text{transmissivity} \times \text{pressure gradient}}{}$$

and

$$\text{flow out} = K_T \cdot \frac{(dP)}{(dh)_T} \text{ m/yr.}$$

$$\text{Decompaction rate} = K_B \frac{(dP)}{(dh)_B} - K_T \frac{(dP)}{(dh)_T} \text{ m/yr.}$$

$$\text{If layer thickness is } R \text{ m,} \atop \text{decompaction strain rate} = \left(K_B \frac{(dP)}{(dh)_B} - K_T \frac{(dP)}{(dh)_T} \right) \Big/ R \text{ per yr.}$$

Here the top of the fraction is the change in KdP/dh from base to top, and the denominator is the change in height from base to top, so the fraction can be written as

$$\frac{d(Kdp/dh)}{dh}.$$

In part (b) of Figure 5.32, K does not change with height so that in this special situation, the fraction can be rewritten as

$$K \frac{d(dP/dh)}{dh} \quad \text{or} \quad K \frac{d^2P}{dh^2}.$$

Summary:

$$\dot{e} = \frac{d}{dh}(K\, dP/dh)$$

$$= K \frac{d^2P}{dh^2} \quad \text{if } K \text{ does not change with } h. \tag{5.8}$$

The corresponding statement for flow of heat is

$$\dot{T} = \frac{d}{dh} (J \, dT/dh)$$

$$= J \frac{d^2T}{dh^2} \quad \text{if } J \text{ does not change with } h. \tag{5.9}$$

Here T is temperature, $\dot{T}$ is the rate of change of temperature with time, and J is a factor combining the material's conductivity and specific heat.

Solution Transfer   The purpose of Figure 5.28 was to emphasize the similarity between two processes, the escape of fluid from among grains of fixed size and the escape of grain rims by solution, leaving behind smaller grain cores. If the processes are indeed similar, an equation like (5.8) should apply to part (b) of Figure 5.28 as well as to part (a). If so, what changes in phraseology need to be made?

First, Question 5.38 dealt with the increase in volume of a layer as its fluid content increased; this is simply the opposite of the process in Figure 5.28. Second, Question 5.38 dealt with only one direction (the vertical) whereas Figure 5.28 shows two dimensions changing and is intended to imply the third changing as well; this also is a difference more in detail than in principle and is taken care of below. The essential difference is the nature of the volume-reduction process, but as regards the material's rheology—the manner in which it responds to the stress state imposed—the difference between parts (a) and (b) of Figure 5.28 does not make much difference.

In part (b), we have to imagine the dissolution process going on continuously. This means (i) that an intergranular solvent is present, and (ii) either the solvent is continuously streamed away and replaced with fresher, more undersaturated solvent or, at the other extreme, that the solvent stays motionless and that solute (the material from the grains) diffuses down the little channelways because of the existence "somewhere else" of more of the same solvent in a more undersaturated state. Rocks *do* diminish in volume; Questions 5.32, 5.33, and 5.34 are drawn from real outcrops; hence, although the possibilities in (ii) seem hard to believe, some such process must have occurred. The main respect in which the ideas are probably unrealistic is the static geometry implied; in reality, rocks heave and swell, and shift a bit here and then later, they shift a bit there. The solution channelways are *not* as uniform through space nor as durable from earlier to later time as Figure 5.28(b) suggests. But if the parent grains are quartz, for example, a stream of quartz streams out of the sample in diagram (b) just as a stream of water streams out of it in diagram (a). We have to accept the idea that a sandstone has a transmissivity K_{water} that goes in Equation (5.8) to describe change of water content, but it also has a transmissivity K_{quartz} that goes in the same equation to describe change of quartz content. And movement of quartz is driven by gradients in the pressure of the quartz just as movement of water is driven by gradients in the pressure of the water.

Direction of Shortening and Direction of Escape

The practical example to which our thoughts are currently pegged is the decompacting layer in Figure 5.32, which relates back to the column of sediments and discussion of Figure 5.31. There it was explicitly supposed that the pile is "broad and flat so that escape of water is upward, uncomplicated by migration sideways." We now consider sideways as a possible direction of escape; see Figure 5.33. The figure shows a rather uniform glacial clay, not totally lithified, that has been overlaid by a boulder bed. This is

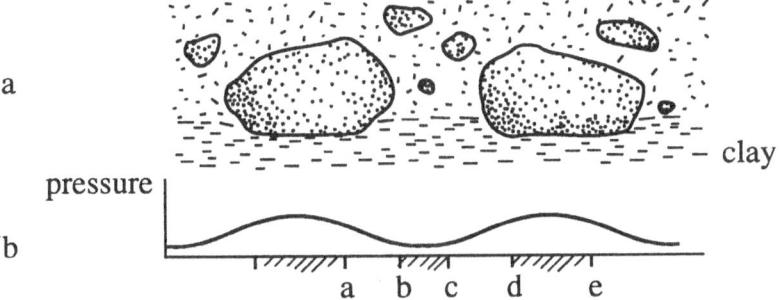

FIGURE 5.33

just one example of a situation where, although water pressure in the clay results from vertical forces (the boulders' weight, plus other overburden), pressure *gradients* are noticeable in a horizontal direction.

In the pressure profile, Figure 5.33(b), there are four distinguishable parts, represented by *ab*, *bc*, *cd*, and *de;* beyond point *e*, the next part simply repeats *ab*. Following the same train of thought as with Figure 5.31 (the vertical pressure profile) we note:

Within *ab*, the gradient is almost uniform; any segment inside *ab* receives flow in from the left and discharges flow out to the right, and any imbalance is small in comparison with the through-flow;
Within *cd*, the situation is similar except that through-flow is in the opposite direction;
Within *bc*, water tends to accumulate; and
Within *de*, water content diminishes as water escapes to either side.

Algebraically, let the horizontal position of a point be designated by a coordinate x and let flow to the right in Figure 5.33 be considered positive. Then we have to put a negative sign into Darcy's law thus:

$$\text{flow rate} = -K \frac{dP}{dx}$$

if we want to keep K, the transmissivity, as a positive number. The rate at which water accumulates within *bc* can be written either as

$$- \frac{(\text{flow to right at } b) - (\text{flow to right at } c)}{x_b - x_c}$$

or as

$$- \frac{(\text{flow to right at } c) - (\text{flow to right at } b)}{x_c - x_b} .$$

By putting x-values into the denominator in the same order as we put flow-rate values into the numerator, we get the same result either way, but following that rule we need a negative sign at the front to get the right overall effect. The lower form in particular can be written as

$$- \frac{\Delta(-K \cdot dP/dx)}{\Delta x} \quad \text{or} \quad \frac{\Delta(K \, dP/dx)}{\Delta x} ,$$

and again if K does not change with x, this can be rewritten with a second derivative as

$$K \frac{d^2 P}{dx^2} .$$

The preceding analysis closely resembles the discussion of Figure 5.31. The difference is that although the loading is still vertical, the flow direction is mainly horizontal; if we designate the vertical direction by Z, Figure 5.31 makes us think about $\partial^2 P_z / \partial z^2$ whereas Figure 5.33 makes us think about $\partial^2 P_z / \partial x^2$ (and of course perpendicular to the plane of Figure 5.33 there might be a $\partial^2 P_z / \partial y^2$ as well). Next we go back to Figure 5.28, where the materials are not layered, and realize that the volume in these samples has *nine* reasons for changing

$$
\begin{array}{lll}
\partial^2 P_x / \partial x^2 & ../\partial y^2 & ../\partial z^2 \\
\partial^2 P_y / \partial\ .\ .^2 & .\ .\ . & .\ .\ . \\
\partial^2 P_z / \partial\ .\ .^2 & .\ .\ . & .\ .\ .
\end{array}
$$

Clearly, a fog of algebraic complexity lies ahead, but it is all of the same type as Equation (5.8). The simplicity of this equation and the physics underlying it—that, as Darcy emphasized, a pressure gradient is apt to drive a flow—the simplicity of the underlying ideas need not be lost sight of.

Special Cases In Equation (5.8) the left-hand side is a strain rate for just the vertical dimension, $\dot{e}_z$. The nine terms just recognized contribute to the *total* rate of change of volume of the sample, $\dot{e}_x + \dot{e}_y + \dot{e}_z$, as in Equation (2.12). Thus,

volume strain rate, $\dot{e}_{vol}$ = sum of 3 linear strain rates
= sum of 9 second derivatives.

For the first special case, let $P_x = P_y = P_z$ i.e., assume that the pressure driving the flow is hydrostatic at every point. Then

$$
\dot{e}_{vol} = 3K \left\{ \frac{\partial^2 P}{\partial x^2} + \frac{\partial^2 P}{\partial y^2} + \frac{\partial^2 P}{\partial z^2} \right\} \quad \text{or} \quad 3K \cdot \nabla^2 P. \tag{5.10}
$$

For the second special case, assume *also* that the change of volume is spherically symmetrical, as with a bubble shrinking or expanding, $\dot{e}_x = \dot{e}_y = \dot{e}_z$. Then

$$
\dot{e}_{radial} = K \left\{ \frac{\partial^2 P}{\partial x^2} + \frac{\partial^2 P}{\partial y^2} + \frac{\partial^2 P}{\partial z^2} \right\}. \tag{5.11}
$$

If we consider a porous sandstone where the quartz does *not* migrate but the pore fluid does, and compressive stresses at some point are σ_x and σ_y, a possible combination of effects is (temporarily ignoring z):

change of shape: $\dot{e}_x^{shape} = (\sigma_x - \sigma_y)/4N$
$\dot{e}_y^{shape} = -(\sigma_x - \sigma_y)/4N$

change of volume: $\dot{e}_{radial}^{vol} = K \left(\frac{\partial^2 P}{\partial x^2} + \frac{\partial^2 P}{\partial y^2} \right)$

giving

$$
\dot{e}_x^{total} = \frac{\sigma_x - \sigma_y}{4N} + K \left(\frac{\partial^2 P}{\partial x^2} + \frac{\partial^2 P}{\partial y^2} \right) \tag{5.12}
$$

and similarly for $\dot{e}_y^{total}$. This is a sum of separate effects of the type already discussed in Chapter 2, in the section Change of Area, Change of Shape. In comparison with the vague statement there that "Q is controlled by one set of processes and R is controlled by another" (where Q is the change-of-shape factor $(1/2)(e_{max} - e_{min})$ and R is the change-of-volume factor $(1/2)(e_{max} + e_{min})$, we now see how the rheological properties N and K enter the quantitative description of the controls.

If, instead, we consider the well compacted sandstone in Figure 5.28(b), we might suppose the mean stress $\bar{\sigma}$ or $(1/2)(\sigma_{max} + \sigma_{min})$ plays a role in place of the water pressure P just discussed. In such a case, we would consider the behavior

$$\dot{e}_x^{total} = \frac{\sigma_x - \sigma_y}{4N} + K\left\{ \frac{\partial^2 \bar{\sigma}}{\partial x^2} + \frac{\partial^2 \bar{\sigma}}{\partial y^2} \right\} \tag{5.13a}$$

$$\dot{e}_y^{total} = \frac{-(\sigma_x - \sigma_y)}{4N} + \text{same.} \tag{5.13b}$$

The role of the material-migration terms, on the right of Equation (5.10) and (5.11), has long been recognized in studies of unconsolidated compaction, and is being increasingly considered as a factor in "solid rock" deformation.

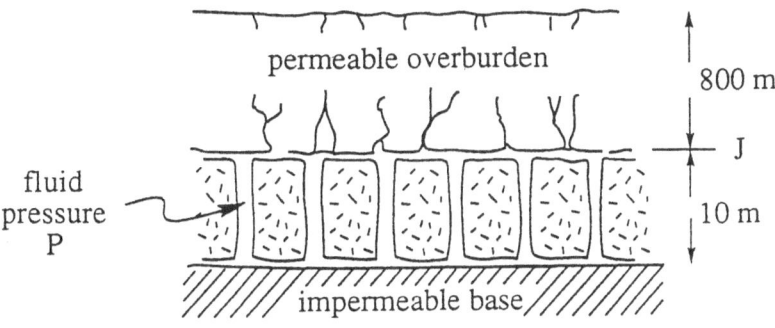

FIGURE 5.34

Question 5.39 Compaction and escape of fluid: A porous sandstone bed 10 m thick is shown idealized in Figure 5.34. We take a two-dimensional approach and assume that, perpendicular to the page, the geometry continues as shown; i.e., long bars of rock are separated by long sheets of fluid. At what rate does the sandstone bed collapse?

Properties: The sandstone layer is 80 percent rock and 20 percent water; that is, the bars are four times as wide as the fluid sheets. The rock bars have viscosity 35 gvu. The overburden has density 2.5 g/cm³ and transmissivity 100 m²/MPa-gtu. There are three unknowns—the vertical compressive stress in the rock bars, the fluid pressure and the rate at which the rock bars shorten vertically. We seek an answer by trial and error, using three pieces of reasoning as follows:

(a) If, as a first trial, we guess that the vertical stress in the rock is 22 MPa, what is the fluid pressure? (Write a force balance at level J for support of the overburden.)

(b) Assume the rock sustains 22 MPa vertically and is confined by just the fluid pressure horizontally.
(i) What is its strain rate? (ii) At what rate does surface J descend, in m/gtu? (iii) Consider a representative horizontal area such as 40 m²: how many cubic meters of fluid have to escape upward per gtu? (iv) What Darcy velocity of the fluid is needed, in m/gtu, to allow the fluid to get away?

(c) For the fluid pressure you calculated, (i) what is the pressure gradient in the overburden in MPa/m? (Assume the gradient is uniform and that the fluid pressure is 0 MPa at the surface.) (ii) By how much does the pressure gradient exceed a hydrostatic gradient? (iii) What Darcy velocity through the overburden will this pressure gradient drive?

(d) Your answer at (b.iv) and (c.iii) probably do not match. (The latter should be about 3/4 × the former.) To get a better match, do you need a smaller or a larger fluid pressure? Do you need to begin with a smaller or a larger trial value for the vertical stress in the rock?

There must be an algebraic method for reaching an exact answer to the problem, but it is more instructive to try some other possibilities arithmetically. To go to a fluid pressure of 15 MPa overshoots the condition sought, and a fluid pressure of 13 MPa comes close to giving the desired match between (b.iv) and (c.iii).

As usual, the purpose of the exercise is not to discover what the stress magnitudes would be but to review the reasoning process and see how the physical quantities affect each other. The fluid pressure adjusts itself until the rate at which the pore space collapses is equal to the rate at which the permeability allows the fluid to escape. The situation can be made more realistic geometrically (and consequently more complicated), but the same basic idea continues to be the key.

Summary

If a sample of rock, such as a few cubic meters, deforms, the deformation can be thought of as having two parts—a change of shape at constant volume and a change of volume at constant shape; that is, a shrinkage or expansion that is the same along all directions in the sample. At any moment, each process goes on at a certain *rate*. In a Mohr circle diagram, the radius of the circle shows the rate of change of shape and the distance of the center from the origin shows the rate of change in volume.

In some circumstances, two separate mechanisms produce the two effects. For example, in deformation of a bag of wet sand, change of shape might occur mainly by the grains sliding past each other, whereas change of volume might occur by water seeping out through the fabric of the bag.

In other circumstances, the *same* mechanism produces the two effects. For example, in compaction of a broad horizontal layer of mud, the thickness might diminish by 10 percent. Spheres initially present in the mud would become ellipsoids and, at the same time, volume would diminish by 10 percent.

Change of volume of a sample depends on the presence of a source or sink for the extra material, and the rate of change of volume will be faster when the source or sink is closer.

Change of volume may involve a source *and* a sink; a sample may be gaining some material (e.g., quartz) and losing other material (e.g., calcite) simultaneously. Change of volume requires only that the rate of gain and rate of loss be different.

Quantitative expressions are of the form

rate of movement = (permeability) × (driving pressure gradient)
    of material

If pressure changes along some direction x, the pressure gradient at site S can be written $(\partial P / \partial x)_S$ and the rate of movement can be written $K_S \cdot (\partial P / \partial x)_S$ m/sec. Then if at a second site T, the rate of movement is $K_T (\partial P / \partial x)_T$ the rate of change of volume between the sites is $K_S \cdot (\partial P / \partial x)_S - K_T (\partial P / \partial x)_T$.

In the special case where K_S and K_T are the same (i.e., the material is of uniform permeability), the rate of change of volume depends on the *second* derivative $\partial^2 P / \partial x^2$.

In rate of change of shape at constant volume, expressions arise such as $e = (\sigma_{max} - \sigma_{min})/4 \cdot (\text{viscosity})$. Where change of shape and change of volume occur together, expressions take the general form

$$\text{strain rate} = \underbrace{\frac{\sigma_1 - \sigma_2}{4 \cdot (\text{viscosity})}}_{\substack{\text{change of} \\ \text{shape}}} + \underbrace{\frac{\partial}{\partial x}\left[K \frac{\partial P}{\partial x} \right]}_{\substack{\text{change-of-} \\ \text{volume terms}}} + \cdots$$

Even where movement of material is mainly horizontal, change of a sample's dimension may be mainly vertical. In general, the direction of material transport and the direction of a sample's shortening or swelling need not be the same.

Large pressure gradients exist mainly between sites that are close together; a sample can change volume rapidly if there is a sink nearby but can change volume only slowly if the sink is remote. Hence lithosphere plates can be considered to deform at constant volume whereas, for hand specimens and outcrops, change of volume should not be ignored.

ANSWERS

Answer 5.1

$$N = \frac{100 \text{ MPa}}{0.1/\text{my}} = 1000 \text{ MPa} - \text{my}.$$

There is some merit in maintaining simple connections with the SI system of units, including time in *seconds*, and this makes 3 1/3 million years ($\cong 10^{14}$ sec) more convenient than 1 million. If we think of 1 MPa as a geological stress unit and 10^{14} sec ($\cong$ 3 1/3 my) as a **geological time unit, gtu,** then a **geological viscosity unit** = 1 MPa/1 per gtu = 1 **gvu.**

Example: Convert Answer 5.1 to gtu and gvu. The rate of 0.1 per million years $\cong$ 0.33 per gtu, so

$$N = \frac{100 \text{ (MPa)}}{0.33 \text{ (gtu)}} = 300 \text{ gvu.}$$

More exactly, 0.1 per million years = 0.3168 per gtu. But the approximation 1 year $\cong$ 3.10^7 sec is already in common use and to be consistent we use

10 my $\cong$ 3 gtu   and   1 gtu $\cong$ 3 1/3 my.

These yield a rate of 0.33 per gtu rather than 0.3168.

Answer 5.2

$$\text{Rate} = \frac{\text{stress}}{\text{viscosity}} = \frac{80}{30} = 2\tfrac{2}{3} \text{ per gtu.}$$

Total time = 3 gtu, so total strain = 8.

Answer 5.3

Consider a column of rock perpendicular to the layer with cross-section A m^2 as in Figure 3.34. Weight = 2100 A t, force component acting parallel to slope = 210 A t, and tangential stress parallel to slope = 210 t/m^2, = 21 bar = 2.1 MPa.

Hence shear strain rate in the gypsiferous layer =

$$\frac{2.1}{1/1000} = 2100 \text{ per gtu.}$$

Thickness of layer = 9 m, so a strain rate of 1 per gtu would be 9 m of displacement per gtu, and shear strain of 2100 = 18,900 m or 18.9 km/gtu, = 0.6 cm/yr. The total distance and total time are probably realistic, but actual sliding sheets probably move in shorter bursts of higher speed.

(The topic is continued in Question 5.16.)

Answer 5.4

Shear strain rate in shale = 24/30 = 0.8/gtu. Total shear strain in 2 gtu = 1.6.

Similarly, total shear strain in limestone = 0.6 (see Figure 5.35).

(Compare with Answers 2.26 and 2.27.)

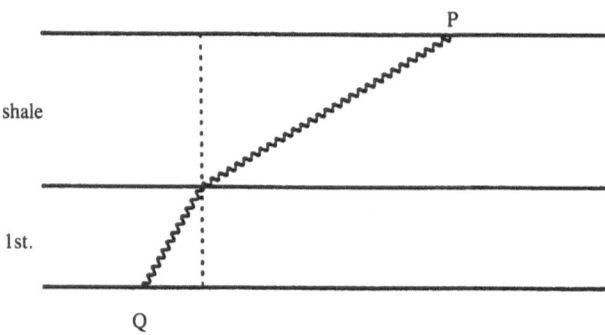

FIGURE 5.35

Answer 5.5

Total sideways displacement of *P* in Figure 5.35 with respect to *Q* = 24 cm. Also vertical separation = 20 cm. Hence

overall slope = 1.2 m per m vertically,
= 1.2 units of shear strain;
angle of shear = 50°.

Overall shear strain $rate = \dfrac{1.2}{2 \text{ gtu}} = 0.6 / \text{gtu}.$

Effective viscosity $= \dfrac{24 \text{ MPa}}{0.6 / \text{gtu}} = 40 \text{ gvu}.$

Answer 5.6

Making the simplest assumptions, shear strain in rock (2) parallel to the interface = 2.5 × shear strain in rock (1). Hence, whatever the time taken for the deformation, (shear strain rate)$_2$ = 2.5 × (shear strain rate)$_1$. Hence, again, whatever the shear-stress magnitude,

(viscosity)$_2$ = (1/2.5) × (viscosity)$_1$.

Answer 5.7

(a) The *most* shear strain would develop if, for the whole of the quiet period, the shear stress was *just under* 40 MPa. This would give shear strain rate = <40 MPa/0.08 gvu = <500/gtu or <0.5 in 1/1000 gtu.

(b) Away from the fault zone, similarly, the maximum possible total shear strain = 0.0008 or about 1/20°. Compare the answers to Question 5.7 with Answer 5.2, where the period of time considered is 3000 times greater. Question and Answer 5.2 could well describe the state of affairs *beneath* the Altyn Tagh Fault, at say 20 km down, where deformation is more steady (with no jerks or earthquakes). At such a depth, there is simply less *contrast* between the central core of the shear zone and its marginal parts.

Answer 5.8

(a)
(i) 100 km of displacement distributed uniformly over 20 km width gives a shear strain rate = 5/gtu [Figure 5.36(a)].

(ii) The same total distributed uniformly over 200 km width gives a shear strain rate = 0.5/gtu.
(iii) The same total distributed entirely within 200 km, but climbing to a maximum at the center

could give a max shear strain rate = 2/gtu (Figure 5.36(b).

(b) Suppose shear stress at 30 km depth *equals* the mean value at the surface, 30 MPa: then the viscosity in the shear-zone center at (iii) equals

$$\frac{\text{stress}}{\text{strain rate}} = \frac{30 \text{ MPa}}{2/\text{gtu}} = 15 \text{ gvu.}$$

Similarly, higher up, where there is only half as much creep, viscosity could equal 30 gvu at the same shear stress; and at the surface, viscosity could equal 150 gvu.

But the shear-stress magnitude is not necessarily the same at all levels: 50 MPa and viscosity 25 gvu at 30 km depth would fit the geometrical requirements, or 10 MPa and 5 gvu. These are major genuine unknowns in earthquake science. The exercise shows how the geometrical requirement of 100 km/gtu can be satisfied in many ways, all following the basic relation: shear strain rate = (shear stress)/viscosity.

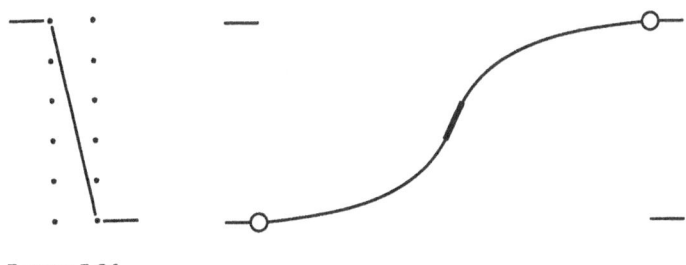

FIGURE 5.36

Answer 5.9

Strain rate $\frac{1}{6}$ per gtu = $\frac{80 \text{ MPa}}{4 \times (?) \text{ gvu}}$;

hence viscosity = 120 gvu.

(The actual motion was probably more spasmodic,

with shorter periods of higher strain rate, perhaps 10 times faster. This would involve the rocks locally and temporarily acquiring lower viscosity such as 12 gvu, for example during a time when fluids were active.

(The topic is discussed again in Chapter 7.)

Answer 5.10

(a) Strain rate = $\frac{120 - 84 \text{ MPa}}{4 \times 120 \text{ gvu}}$ = 0.075 per gtu.

(b) Halfwidth = 75 km, giving displacement rate

5 1/2 km/gtu (to the nearest 1/2 km) at the southern end, or about 16 km during the 10 my of activity.

Answer 5.11

TABLE 5.3

rock viscosity	stress difference 100	30	10	3 MPa
100 gvu	1/4 per gtu	3/40	1/40	—
10 gvu	2 1/2	3/4	1/4	—
1 gvu	25	7 1/2	2 1/2	3/4

Answer 5.12

First solution: Focus on stress magnitudes. To give $\tau_{max} = 80$ MPa, we need $\sigma_{max} - \sigma_{min} = 160$ MPa. Then shortening strain rate = 160 MPa/4 × 30 gvu = 1 1/3 per gtu.
Second solution: Focus on strain rates. If the *shear*

strain rate that develops is 2 2/3 per gtu, then in the condition $\dot{e}_{max} = -\dot{e}_{min}$, we need $\dot{e}_{max} = 1$ 1/3 per gtu, because of the requirement $\dot{e}_{max} - \dot{e}_{min} = \dot{\gamma}_{max}$.

Answer 5.13

Strain rate in shale $= \dfrac{60 - 40 \text{ MPa}}{4 \times 30 \text{ gvu}} = \dfrac{1}{6}$ per gtu.

For strain rate in limestone to be the same, we need

$$\dfrac{60 - (?) \text{ MPa}}{4 \times \text{viscosity}} = \dfrac{1}{6} \text{ per gtu.}$$

Hence (?) = 6²/₃ MPa.

Note: keeping the stress *differences* in focus and their actual magnitudes in the background, we see that the stress differences are proportional to the rocks' viscosities

$$\dfrac{\text{(stress difference) shale}}{\text{(stress difference) limestone}} = \dfrac{3}{8} = \dfrac{\text{(viscosity) shale}}{\text{(viscosity) limestone}}.$$

Answer 5.14

68 MPa in limestone; 63 MPa in shale.

Answer 5.15

(stress difference)$_{limestone}$ = 24 MPa; joint strain rate = .075 per gtu.

And so on.

Answer 5.16

For shear strain of 0.3, we need displacement = 0.3 × 9 m = 2.7 m. At 0.6 cm/yr, the time needed is 450 years. Then viscosity drops by 1/3× so displacement rate increases 3×, to 1.8 cm/yr, and time needed for the next 0.3 units of shear strain = 150 years; and so for the third unit, 50 years.

Note: Although the stepwise aspect of the calculation

is unrealistic, the overall effect is not. Down-slope movements often *are* self-accelerating, and the behavior can be described by a viscosity that is diminished by strain. To be more realistic, the main change needed would be to have a larger number of smaller increments (a type of calculation for which electronic calculators are particularly well suited).

Answer 5.17

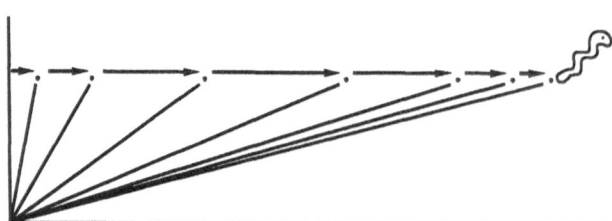

FIGURE 5.37

Answer 5.18

The stress state in Figure 5.13 is simply a combination of the stress states discussed in Questions 5.4 and 5.13. Strain rates are: along p and along r, shortening at $-1/6$ per gtu; along q and s, elongation at $1/6$ per gtu; shear strain rate of the pair pq is 0.8 per gtu and shear strain rate of the pair rs is 0.3 per gtu. After 1 gtu, the outline would be approximately as shown in Figure 5.38.

Note: It is not so easy to say *exactly* what the shape would be after 1 gtu; see Question 2.6 through Equation (2.4). For linear change, exact statements are

$$\log \text{ (stretch along } p) = -1/6$$
$$\log \text{ (stretch along } q) = \quad 1/6$$

$\log [(\text{stretch along } p) \times (\text{stretch along } q)] = 0$
$(\text{stretch along } p) \times (\text{stretch along } q) = 1$

as required if area is to be exactly conserved.

For shear strain, refer to Figure 2.27: if history (a) were followed, the shear strain would be exactly 0.8 in the shale; if history (b) were followed, the shear strain would be slightly more than 0.8 because the angles of the parallelogram would be changed by the linear strain *after* 0.8 of shear strain had been introduced; and history (c), which is of the type where shear strain continuously accompanies linear strain as in our problem, is intermediate between (a) and (b). Shear strain in history (c) thus cannot be 0.8 *exactly* because a little of effect (b) creeps in. As usual, the statements about strain rates are simple and exact, but statements relating finite changes to a material's properties require more care. The previous section emphasized that a rock's material properties tend to change during deformation, so that the strain that develops over a finite time usually cannot be related to a stress field exactly. An approximate picture as in Figure 5.38 is usually as much as can be justified.

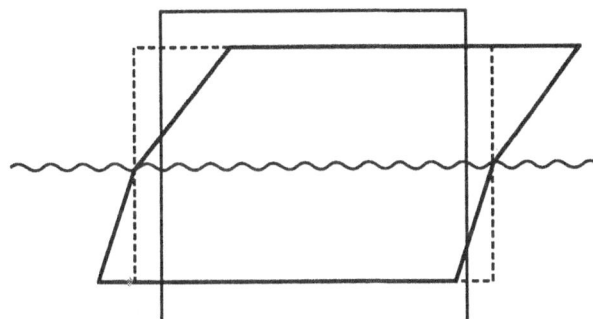

FIGURE 5.38

Answer 5.19

The shear angle α is part of a triangle with sides h and $0.8\,h$ as shown in Figure 5.39. The effect of the normal stresses is to shorten h by a strain of about $-1/6$, and to elongate $0.8\,h$ by a strain of about $+1/6$. Tan α was formerly 0.8 and becomes 0.93/0.83 or 1.12; the new value of tan α is 1.4 times the original value.

Now consider the limestone: shear angle β is part of a triangle with sides k and $0.3\,k$. Following the same train of thought gives new dimensions $0.83\,k$ and $0.35\,k$, and a new tangent 0.35/0.83 or 0.42. Again the new value is 1.4 times the original value. The ratio tan α/tan β does not change at all.

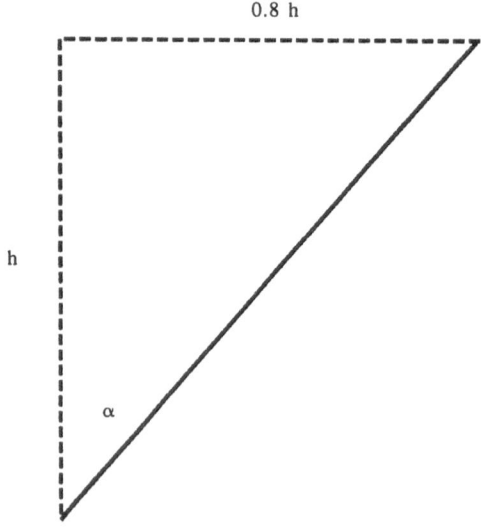

0.8 h

h

α

FIGURE 5.39

Answer 5.20

The original shape being a square, we can imagine a circle drawn inside it. The upper semicircle becomes a semi-ellipse that touches the boundaries at points p, q, and r; the lower semi-ellipse touches at p, r, and s (q and s being midpoints). Once the ellipses are drawn,

their long axes are found by inspection (see Figure 5.40).

Note: To get an exact answer, one would use matrix algebra.

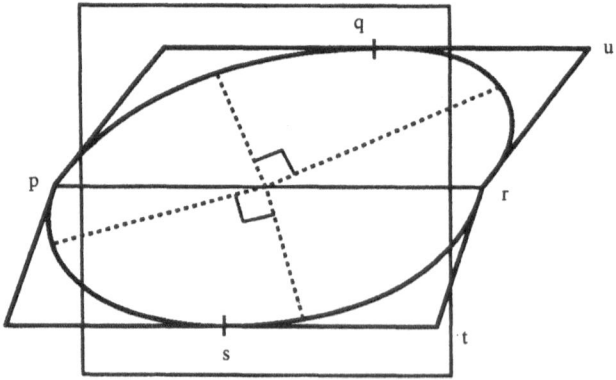

q

u

p

r

s

t

FIGURE 5.40

Answer 5.21

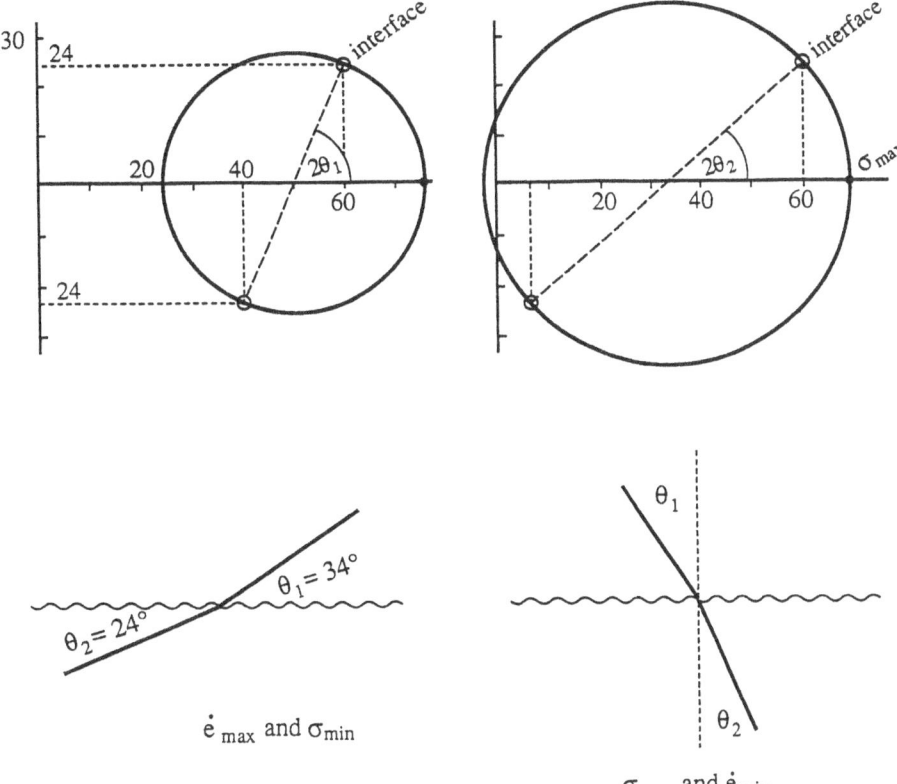

FIGURE 5.41

Answer 5.22

The thicknesses show that, in *fractions*, the succession consists of 3/5 shale and 2/5 limestone. The weighted mean stress parallel to the layering is therefore $(3/5) \cdot 40 + (2/5) \cdot 6\ 2/3$ or $26\ 2/3$ MPa. With compression normal to the layering of 60 MPa and overall linear strain rate 1/6, the effective viscosity is $(60 - 26\ 2/3)/4 \cdot (1/6)$ or 50 gvu.

Answer 5.23

Consider first flattening, driven by a normal stress of S MPa acting on layers with viscosities M and N that form fractions h and k of the assembly ($h + k = 1$); see Figure 5.42. If the compressive stresses parallel to the layers are P and Q and the linear strain rate is e everywhere in the assembly, then

$$e = \frac{S - P}{4M} = \frac{S - Q}{4N}$$

or $P = S - 4\,Me$ and $Q = S - 4\,Ne$.

Then the effective stress on the sides, $hP + kQ$, is

$hS - 4\,hMe + kS - 4\,kNe$ or $S - 4e(hM + kN)$.

If we name this effective stress R, the effective viscosity of the assembly is $(S - R)/4e$, which works out to be $hM + kN$.

Consider next layer-parallel shearing. In this situation, the same shear stress affects all the layers; let us name it T (MPa). Then shear strain rates are T/M and T/N per gtu. If we consider a thickness of shale of H cm and keep its base stationary, the displacement rate of its top surface will be HT/M cm/gtu. For a limestone layer of thickness K cm, the corresponding rate is KT/N cm/gtu, and if the limestone rides on the shale, the total displacement rate will be $HT/M + KT/N$ or $T(H/M + K/N)$. The total thickness is $H + K$ cm, so that the overall or effective shear strain rate is $T(H/M + K/N)/(H + K)$. But $H/(H + K)$ and $K/(H + K)$ are the thickness fractions h and k, so that the effective shear strain rate is $T(h/M + k/N)$. Now viscosity = (shear stress)/(shear strain rate) so that the effective

viscosity of the assembly is $T/T(h/M + k/N)$ or $1/(h/M + k/N)$

$$\frac{1}{\text{effective viscosity}} = h\frac{1}{M} + k\frac{1}{N}.$$

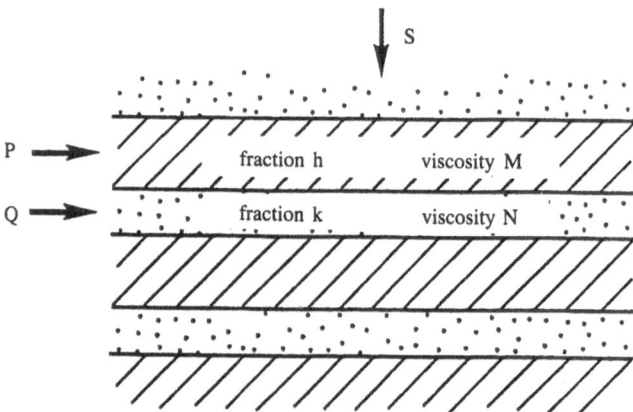

FIGURE 5.42

Answer 5.24

The shear stress of 24 MPa, treated as in Figure 5.15, would produce a shear strain rate of 24 MPa/40 gvu or 0.6 per gtu. The compressive stresses of 60 and 40 MPa would produce linear strain rates of 20 MPa/(4 × 50 gvu) or 0.1 per gtu, (+0.1 per gtu in one direction and −0.1 gtu in the other).

Answer 5.25

The methods of Chapter 4, either graphical and algebraic, give

compressive stress normal to schistosity	385.6 MPa
compressive stress parallel to schistosity	374.4 MPa
stress difference σ(normal) − σ(parallel)	11.2 MPa
shear stress parallel to schistosity	19.2 MPa

Then linear strain rates are 11.2/4 × (viscosity for linear strains parallel to schistosity) or 0.023 per gtu (shortening normal to schistosity and elongating parallel to it). The shear strain rate is 19.2/(viscosity for shear strains parallel to schistosity) or 0.64.

Answer 5.26

See Figure 5.43; $2\theta = 72°$, $\theta = 36°$. These values are not greatly different, but are perceptibly different, from the corresponding angles in Answer 5.21 (Figure 5.41).

Answer 5.27

In Figure 5.43, the shear strain rate plotted is six times the liner strain rate. In the schist, the ratio is 27 times, so that 2θ must be almost 90° (more exactly, 86°). Hence θ is 43.°

The angle α in Figure 5.16 is 37° and is the angle between σ_{min} and the schistosity. Thus $\dot{e}_{max}$ and σ_{min} differ in direction by 6.°

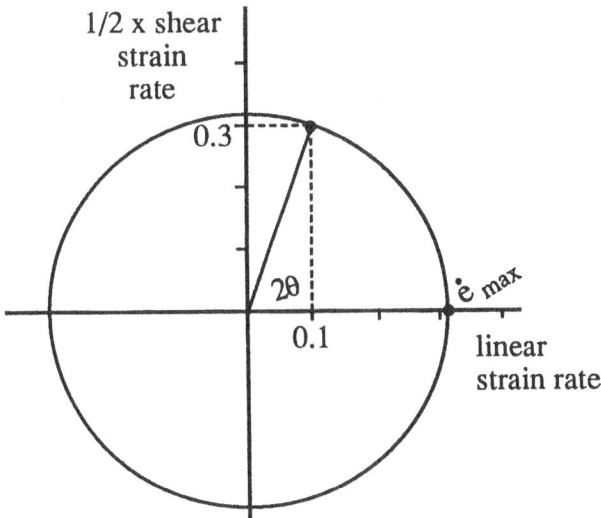

FIGURE 5.43

Answer 5.28

(a) (3 m) × (.023 per gtu) = 0.069 m/gtu,
   (4 m) × (.023 per gtu) = 0.092 m/gtu,
   (4 m) × (0.64 per gtu) = 2.56 m/gtu.
(b) (4/5) × 0.069 = 0.055 m/gtu to left,
   (3/5) × 0.092 = 0.055 m/gtu to left,

(4/5) × 2.56 = 2.05 m/gtu to right.
(c) Net horizontal velocity = 2.05 − 0.055 − 0.055
   = 1.94 m/gtu.
(d) *RS* has length 5 m, so rotation rate is 1.94/5 or
   0.39 radians per gtu or about 23° per gtu.

Answer 5.29

(a) (3/5) × 0.069 = 0.041 m/gtu down,
   (4/5) × 0.092 = 0.074 m/gtu up,
   (3/5) × 2.56 = 1.54 m/gtu up;
   Total vertical velocity = 1.57 m/gtu.
(b) Linear strain rate = 1.57/5 = 0.31 per gtu.
   Viscosity = (40 MPa)/4 × (0.31 per gtu)
   = 32 gvu.

It is not surprising to find that the effective viscosity of
the assembly for horizontal shortening is close to the
effective viscosity of the contained material for shearing (30 gvu), because when $\alpha = 37,°$ shearing is the
predominant process, although not the only process,
going on.

Answer 5.30

$$(400 − S)/4 × \text{(viscosity)} = 0.31 \text{ per gtu.}$$

In the center section, viscosity = 120 gvu, giving S = 251 MPa.

Answer 5.31

(a) $\dfrac{\tan 38°}{\tan 22°} = \dfrac{0.78}{0.40} \cong 2.$
(b) As discussed following Question 5.19, a likely
   source of error is that the two rock types might
   suffer different amounts of linear strain normal to
   the layering by solution processes. Material from
   the limbs may migrate to fold crests to form saddle
   reefs as just discussed, or it may migrate to some
   more distant location. Either way, there is no rea-

son to suppose that this process will affect the two
rock types equally or proportionately.
(c) If the schistosity was parallel to the layering, the
   ratio estimated would relate to the *minimum*
   viscosity of the schist. The schist's maximum viscosity would affect the stress pattern, rate of fold
   closure, and so on but would not affect the ratio of
   tangents of apparent shear angles.

Answer 5.32

(a) In Figure 5.25, the *tallest* spike in each profile is the best indicator of the amount dissolved: 4 mm + 7 mm + 7 mm = 18 mm lost from the three layers. Thickness remaining = 3 slabs like the ones shown = 45 mm (estimated). Hence ratio requested = 18/45 or 0.4; also original extent = 63 mm, so shortening strain = 18/63 = 0.29.

(b) If all measurements are made along the same line, roughly normal to the dark seams, proportions are maintained and no error is caused.

Answer 5.33

In the present mica-poor laminae, 12 mm³ of mica are spread through 100 mm³ of rock; so 72 mm³ of mica would be spread through 600 mm³ of rock.

At present, in mica-rich laminae, 72 mm³ of mica are spread through 100 mm³ of rock; i.e., present vol = 1/6 × original vol, and present thickness = 1/6 × original thickness.

Present thickness = 1.8 mm, so original thickness = 10.8 mm and thickness lost = 9.0 mm.

Putting a pair of laminae together (one mica-rich + one mica-poor), total thickness originally was 10.8 + 3.2 = 14 mm.

$$\frac{\text{Thickness lost}}{\text{Thickness originally present}} = \frac{9}{14}$$

$$= 0.64 \quad \text{(almost 2/3 gone);}$$

$$\frac{\text{volume lost}}{\text{volume originally present}} = \text{same.}$$

Answer 5.34

The greatest and least numbers are 10 per cm and 7 per cm, compared with 5 per cm in the undeformed example, diagram (a). The stretch is 0.5 in one case and 0.7 in the other; we can take the directions to be roughly at 90° to each other, so that the "area stretch" (final area/original area) = 0.5 × 0.7 = 0.35, as in equation (2.6).

Answer 5.35

To use the relation

$$\text{Darcy velocity} = 12.10^{-8} \times \text{(H)}$$
$$\text{m/sec} \qquad \text{m}^2/\text{Pa-sec} \qquad \text{Pa/m}$$

we need to estimate the pressure gradient H as follows.

Pressure at top side of plug = 0.4 bar = 0.04 MPa;

Pressure at lower side = negligible;

$$\text{Pressure gradient} = \frac{0.04 \text{ MPa}}{5 \text{ cm}} = 0.8 \text{ MPa/m}$$

$$= 8.10^5 \text{ Pa/m;}$$

so Darcy velocity = $12.10^{-8} \times 8.10^5 = 0.096$
   m²/Pa-sec   Pa/m   m/sec

$$\cong 10 \text{ cm/sec.}$$

An alternative pair of correct statements is as follows.

Pressure at top side of plug = 0.4 bar + atmospheric pressure;

pressure at lower side $\cong$ pressure at the lower open end, which is atmospheric pressure.

The same value as before would be calculated for the pressure gradient.

Answer 5.36

A pressure at *M* that is 0.04 MPa greater than at *N* is exactly the pressure you calculated in Question 5.35, from the weight of the 4 m of water. That is, this much pressure at *M* just *supports* the water column, without driving any upward flow.

Answer 5.37

(a) If the pipe lay horizontally,

$$\text{Darcy velocity} = 12 \cdot 10^{-8} \times \left[\frac{0.04 \text{ MPa}}{4 \text{ m}} \right]$$

$$\underset{\substack{\text{m}^2/\text{Pa-sec} \\ \text{transmissivity}}}{} \quad \underset{\substack{\text{pressure gradient} \\ = 10^4 \text{ Pa/m}}}{}$$

So Darcy velocity $= 12.10^{-4}$ m/sec $= 1.2$ mm/sec.

(b) 1.2 mm/sec   (c) 0.08 MPa

The overall idea is that, in standing water, vertical pressure gradient is 0.01 MPa/m, or 10 MPa/km as in Answer 3.5; i.e., 0.01 MPa/m just keeps the water standing. It is *deviations from* that hydrostatic situation that drive the flow.

Answer 5.38

Left-hand picture:
Flow in at base =
$\quad (18.10^{-8}) \times (40) \quad$ or $\quad (18.10^{-8}) \times (0.04)$
$\quad$ m²/MPa-yr   MPa/km $\qquad\qquad\qquad$ MPa/m
$\qquad\qquad\qquad = 72.10^{-10}$ m/yr;

Flow out at top $= (12.10^{-8}) \times$ same $= 48.10^{-10}$
$\quad$ m/yr.
Layer must be increasing in thickness at 24.10^{-10}
$\quad$ m/yr.
Original thickness (4 cm) $= 4.10^{-2}$ m.
Strain rate $= 6.10^{-8}$ per yr.

Right-hand picture:
By a similar process, flow in at base $= (35.2)10^{-10}$
$\quad$ m/yr;
Flow out at top $= (22.4)10^{-10}$ m/y.
Layer must be increasing in thickness at 13.10^{-10}
$\quad$ m/yr.

Answers:
The left-hand layer is swelling faster, at 0.24 cm/my.
Strain rate $= 0.06$ per million yr $= 0.2$ per gtu.

Answer 5.39

(a) Overburden pressure $= 20$ MPa; fluid pressure $= 12$ MPa; check, $(0.8) \times 22 + (0.2) \times 12 = 20$ as needed.
(b)
$\quad$ (i) Strain rate $= (\sigma_1 - \sigma_2)/4N = 0.07$ per gtu.
$\quad$ (ii) 0.7 m/gtu.
$\quad$ (iii) 28 m³ of fluid have to escape per gtu through 40 m² of horizontal area.
$\quad$ (iv) 0.7 m/gtu, the same as at (ii).

(c)
$\quad$ (i) 0.015 MPa/m.
$\quad$ (ii) 0.005 MPa/m (hydrostatic gradient $= 0.01$ MPa/m).
$\quad$ (iii) Darcy velocity $= 0.5$ m/gtu.
(d) A *larger* fluid pressure would give a larger rate at (c.iii) and a smaller rate at (b.iv) so that a *smaller* rock stress than 22 MPa is needed.

Parting

An earthquake is, first and foremost, a disaster; lives are cut short, hopes and pleasures vanish. People who have never experienced an earthquake are fortunate; it is something of a luxury to consider the fracturing of the earth from only an academic point of view.

Among their consequences, earthquakes often leave broken and offset fences. These embody most clearly the essence of the present chapter, and the difference from Chapter 5: something that was geometrically continuous no longer is so. A comparable example appears in Figure 5.8, where the topic is boudinage, with coherent and incoherent interfaces. Where lines remain continuous, no matter how much distorted, Chapter 5 applies; it is when discontinuity appears—when something that was formerly continuous parts into two pieces—that we need the present chapter.

TWO KINDS OF PARTING: THE FAILURE ENVELOPE

Two different ways in which a material can part are shown in Figure 6.1. As long as the parting surface is planar, these are easy to distinguish. Parting on a nonplanar surface is more complicated, and going down to micro-scale reveals more complexities again, but at least for a start, we suppose that any parting resembles one or the other of these two. They will be called **shear parting** and **tensile parting** although, as already noted on page 120, the role of stresses that are actually tensile may be small or perhaps nonexistent.

Experimental Arrangements

The common arrangements for generating partings in a laboratory are shown in Figure 6.2. A tensile test produces tensile parting and a shear test produces shear parting—very straightforward. A compression test usually produces shear parting but can produce tensile parting or a combination of the two (see Figure 6.3). Details of tests that have been devised should not obscure the fact that there are just *two* basic types of response.

If one has many similar samples of one homogeneous material, one can perform a suite of varied tests, but the results can usually be summarized as in Figure 6.4. All stress states of the dotted-line set produce tensile parting and all stress states of the solid-line set produce shear parting. Hence, properties of the material that summarize its behavior are the magnitude T, the tensile strength, and the position of the **failure envelope,** also called the **Mohr envelope** (see Figure 6.5). (Sometimes the envelope is continued so as to pass through the point T, but the extra portion of curve is rarely of significance in geology.) The envelope not only shows that some particular stress state is likely to cause failure; it also shows the orientation of the plane that is likely to fail and the stresses on that plane (see Figure 6.6).

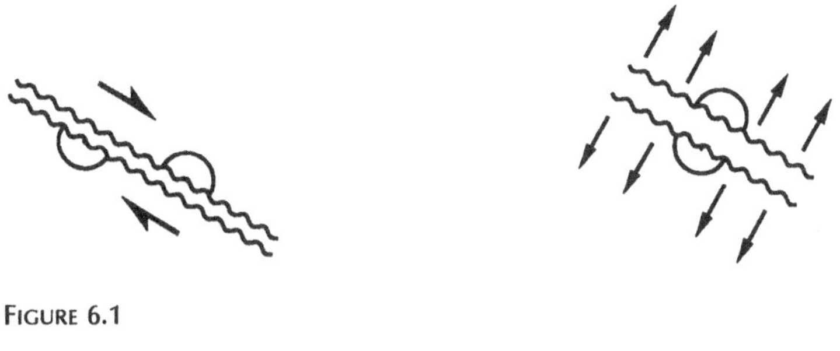

FIGURE 6.1

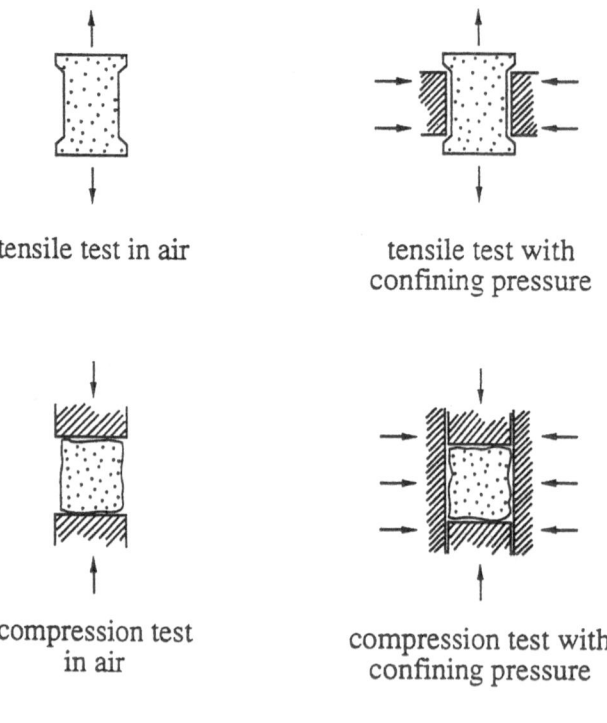

tensile test in air

tensile test with
confining pressure

compression test
in air

compression test with
confining pressure

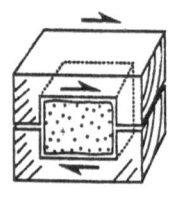

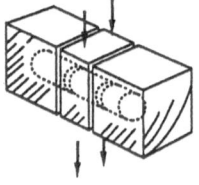

shear tests

FIGURE 6.2

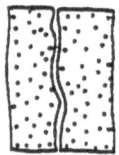

FIGURE 6.3

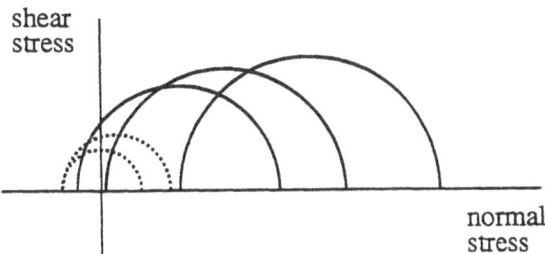

Figure 6.4

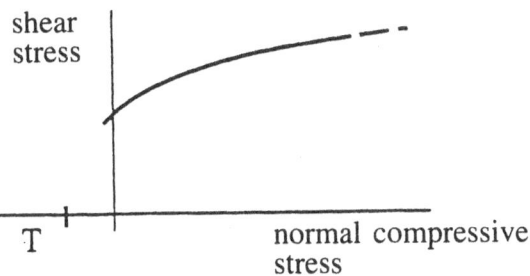

Figure 6.5

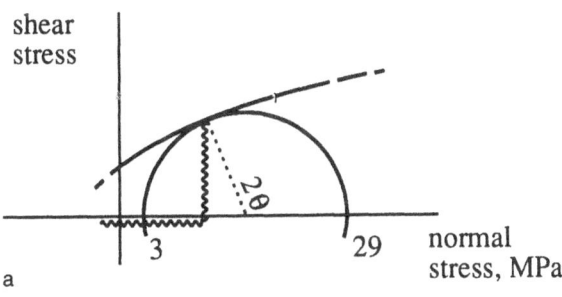

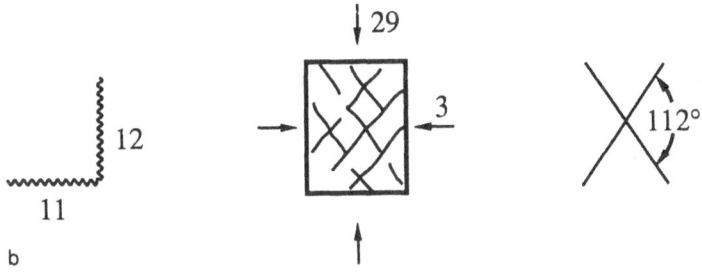

Figure 6.6 For the envelope shown, one of the stress states that gives failure can be specified by $\sigma_{max} = 29$, $\sigma_{min} = 3$ MPa or by $\sigma_{mean} = 16$, $\tau_{max} = 13$ MPa. The two descriptions are equivalent; either specifies the entire circle. We read from the diagram that stresses on the plane that fails are: normal stress 11 MPa; shear stress 12 MPa; $2\theta = 112°$, $\theta = 56°$.

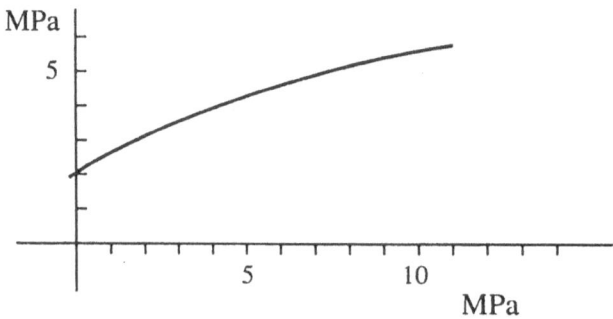

FIGURE 6.7

Question 6.1 Conditions for failure: A material has the failure envelope shown in Figure 6.7.

(a) If σ_{min} = 3 MPa, what value of σ_{max} causes failure?
What is the normal stress on the plane that fails?
What is the shear stress on the plane that fails?
What is the maximum shear stress in the sample?
What is the normal stress on the plane with maximum shear stress?

(b) If σ_{max} = 9 MPa, what value of σ_{min} causes failure?
What is the normal stress on plane that fails?
What is the shear stress on plane that fails?

(c) Draw a pair of fractures making an angle such as might form in the conditions of part (a); repeat for part (b). Does the acute angle between intersecting fractures become more acute as you go to lower pressures or high pressures?

(The topic is continued at Question 6.7.)

Question 6.2 Conditions for failure, continued:

(a) A rock at 400°C fails in the shearing manner at mean stress = 12 MPa, stress difference σ_{max} − σ_{min} = 18 MPa and another piece fails at mean stress 34 MPa, stress difference 28 MPa. Predict a stress difference needed to make the same rock fail at a mean stress of 20 MPa.

Show your answer in a diagram.

(b) The rock fails in tension when tensile stress = 2 MPa. If σ_{max} is set at 12 MPa and σ_{min} is gradually reduced, will the rock fail first in the tensile or the shearing manner?

(c) If σ_{max} is set at 6 MPa and σ_{min} is gradually reduced, same question again.

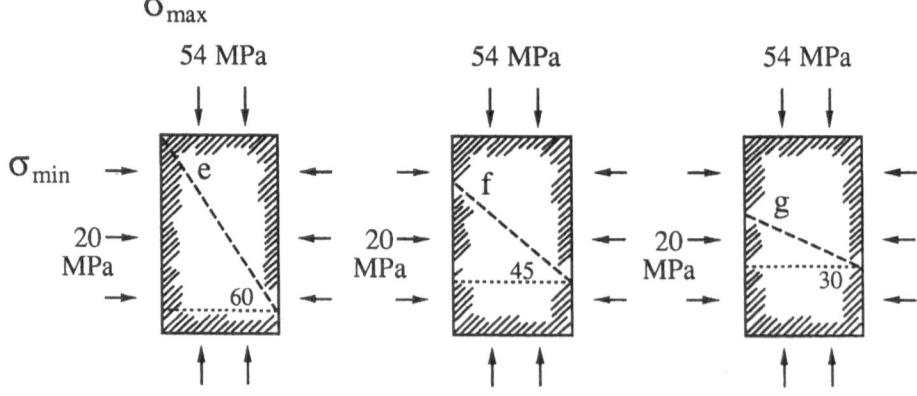

FIGURE 6.8

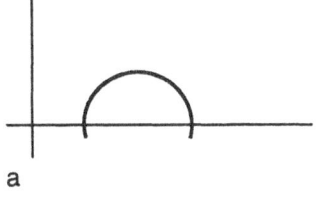 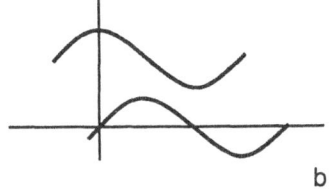

a   b

FIGURE 6.9

Question 6.3 Orientation of plane that fails first: Give estimates of the normal stress and shear stress on planes *e*, *f*, and *g* in Figure 6.8. (Calculate, or estimate from a diagram like Figure 6.9(a) or (b), as in Answer 4.7.)

The shear stress on plane *e* is equal to the shear stress on plane *g*, yet plane *e* is more likely to fail in the shearing manner. Why is this? (Do not refer to the failure envelope; give attention more to the material grains, atoms, and so on.)

Stress Magnitudes

Question 6.4 Geological stress states: Refer to Answer 3.5, showing typical pressures due to overburden in the earth. To understand rock structures as seen in outcrop, we rarely need to consider depths greater than 40 km, and most structures seen formed at depths less than 20 km. Consider a point at 15 km depth where the stress difference $\sigma_{max} - \sigma_{min} = 30$ MPa. Plot a stress circle for such a point. On the same axes, plot points at normal stress 600 MPa, shear stress 100 MPa; normal stress 600 MPa, shear stress zero; and normal stress zero, shear stress 10 MPa. Using these three points and the origin as vertices outline a quadrilateral. Structural geologists mostly operate with stress states within this outline.

Can you show a tensile strength of 3 MPa on the same axes?

For engineering purposes, the order of magnitude of the overall compression is smaller. Few engineering works have as much as 1 km of overburden, or 30 MPa vertical compressive stress. On the other hand, engineers are sometimes able to be selective and to work with rock far more free of fractures than "average geological rock"; and they sometimes wish to contain stresses several times larger than the overburden stress.

Question 6.5 Engineering stress states: On the same axes as in Question 6.4, plot points at normal stress zero, shear stress 40 MPa; normal stress 100 MPa, shear stress 100 MPa; normal stress 100 MPa, shear stress zero. Using these three points and the origin as vertices, outline a second quadrilateral. The diagram now gives a rough pictorial representation of the difference between the stress states an engineer commonly envisages and those of a structural geologist.

Another difference of crucial importance, of course, is the time scale on which each works. Creep deformation at a slow, fairly steady rate may be negligible in an engineering project with life span 100 years while being the dominant effect in a tectonic event with a life span of 10 million years (3 gtu).

Drilling for oil or gas is an exception to the statement just made, and involves engineering commonly under 2 to 5 km of overburden (less commonly under 9 km).

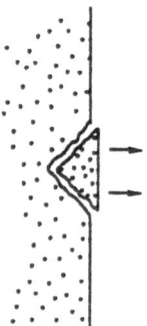

FIGURE 6.10

Question 6.6 Fracture of walls of boreholes:

(a) Suppose you are drilling a borehole through impermeable rocks whose shear strength is 60 MPa under all circumstances. How deep can you expect to drill before the type of failure illustrated in Figure 6.10 sets in? (In this part, neglect the presence of drilling fluid. Assume every 100 m of rock overburden gives 3 MPa of vertical compressive stress. Treat the problem just in the two dimensions shown.)

(b) If you keep the hole filled with drilling fluid, will this type of failure set in before depth 4800 m or not? Explain your decision; assume every 80 m of fluid generates 1 MPa of hydrostatic stress.

ATTITUDE OF FAULTS AND FRACTURES

All the statements made so far about parting have been independent of geographical directions, but the resulting surfaces, if large enough to be called faults, have their own terminology:

1. If movement of one block with respect to the other is horizontal, the fault is a **strike-slip fault.**
2. If the movement of one with respect to the other is up or down the dip of the fault surface, the following terms apply:

	If hanging-wall block moves UP dip:	If hanging-wall block moves DOWN dip:
If dip is shallow:	**thrust**	**slide**
If dip is steep:	**reverse fault** (probably $\sigma_{horiz} > \sigma_{vert}$)	**normal fault** (probably $\sigma_{horiz} < \sigma_{vert}$)

Question 6.7 Thrust and normal faults: At a depth where σ_{vert} is 20 or 21 MPa, what magnitude for σ_{horiz} would initiate a thrust or reverse fault? Use the failure envelope in Answer 6.2.

To what magnitude would σ_{horiz} have to be reduced, to allow normal faulting to occur? (At what depth do these orders of magnitude apply?) What is the magnitude of the normal stress on the fault plane in each case? Consider how much rock-crushing is likely to accompany faulting in the two situations.

Question 6.8 Angle of dip on normal faults: Consider again the normal fault in Question 6.7 under $\sigma_{vert} = 21$ MPa, $\sigma_{horiz} = 3$ MPa. What is its probable angle of dip? If we consider faults of the same type at successively smaller magnitudes of σ_{vert}, what range of dip angles shall we expect? (Question 6.1 is related to this one.)

INHOMOGENEOUS ROCK AND THRUST SHEETS

As already noted, engineers can sometimes arrange to work with "intact rock," whereas a structural geologist works with what nature provides, which is usually a rock containing fractures, or some kind of preexisting weakness. If a fracture is present, the rock's response will be mainly determined by the behavior of the fracture, more than by the behavior of the bulk rock. On a small scale, the bulk properties may be locally significant, e.g., in parting a single horizon into boudins; but in general, deformation effects at weaknesses outweigh deformation effects in the material between weaknesses. This is true

for intergranular slip in unconsolidated sediment,
for grain-boundary slip in low-grade metamorphism,
for any deformation where solution and redeposition is important,
for deformation of layered successions with weak bedding planes,
for fractures rock masses,
and, on the largest scale, for mountain-belt roots shot through with shear zones.

An unfortunate aspect is that, often, the greater the deformation, the more it becomes localized in a very thin layer of rock, and the less conspicuous it becomes among field observables. To approach the mechanics of these important but inconspicuous effects, we consider slip on bedding planes which, in a similar way, can be large in amount but hard to detect.

Slip on Bedding Planes

Figure 6.11 illustrates the difference between three conceivable curves. The lowest is not actually a *failure* envelope, but is of considerable use in structural geology all the same.

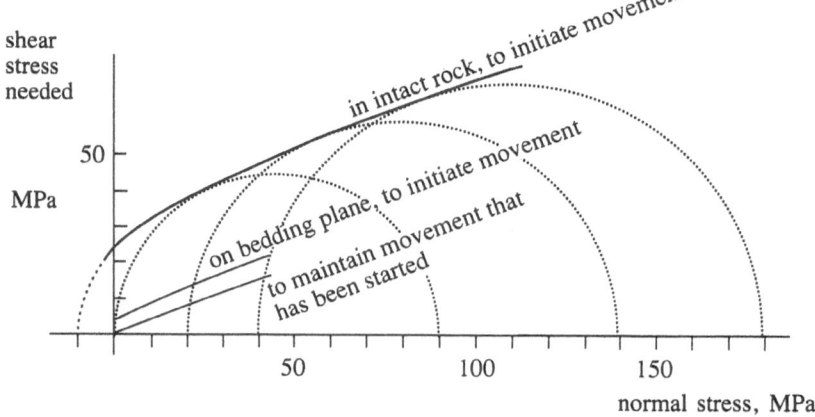

FIGURE 6.11

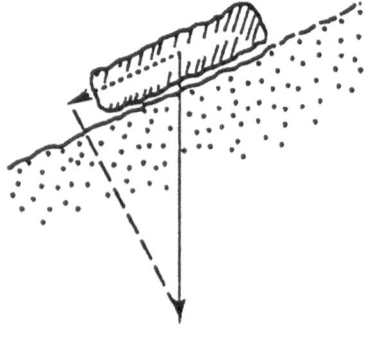

FIGURE 6.12

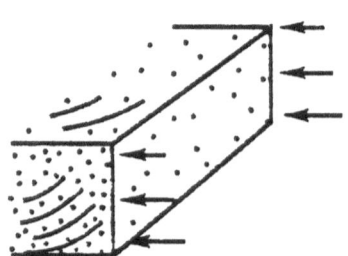

FIGURE 6.13

Question 6.9 Force needed to move a sliding sheet:

(a) Suppose a horizontal sheet or slab of rock 18 km long, 12 km wide, and 1300 m thick is to slide on a bedding plane whose behavior is as in Figure 6.11. What total force is needed (i) to initiate movement (ii) to maintain movement once started? (Express the force in tons where 1 ton of force = weight of 1 ton of rock, and use the approximation 1 MPa = 100 tons/m²; make a realistic assumption about the density of the rock.)

(b) Express each force as a fraction of the slab's weight.

(c) *Check:* In Figure 6.12, to what angle would we need to tilt the slab to generate a force parallel to its base that equalled such a fraction of the slab's weight? (This is simply a check on orders of magnitude. Most slabs slide under their own weight for tilts between 5° and 50° and we should be inside that range.)

(d) What normal stress applied uniformly all over the slab's long edge would generate the smaller force?

If we call this uniform stress σu_{max}, what value of σ_{min} is needed if intact rock is to remain intact and not fail?

(e) In Figure 6.13, can you devise a nonuniform distribution of stress that has a mean value as great as the value in part (d) and does not cause failure in intact rock at any level?

Tapered Slab   The following discussion is intended to show that a tapered slab can be pushed without crushing, where a similar planar slab could not. Just as important, the discussion shows that no new ideas are needed; the mechanical inquiry continues to use the same simple ideas as before. We need perhaps a little more patience, but we know all the necessary mechanics.

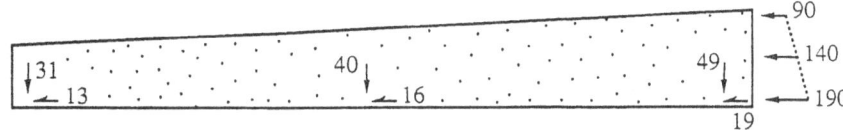

FIGURE 6.14

Let the slab taper from 1600 m at the back end to 1000 m at the front. Instead of a uniform vertical stress all over the base, of 40 MPa, we now have a nonuniform stress, 49 MPa at the back diminishing to 31 MPa at the front. The shear stress needed to maintain motion is now 19 MPa at the back and 13 MPa at the front but the average value and the total force needed are as before.

The problem now is to distribute a total force of 35×10^{10} t over a larger area and avoid crushing the rock. At the surface, where $\sigma_{vert} = 0$, we still have $\sigma_{max\ horiz} = 90$ MPa as before. At the base, we now have $\sigma_{vert} = 49$ MPa and shear stress = 19 MPa, so σ_{max} and σ_{horiz} can go as high as 190 MPa. (Like σ_{vert} and σ_{min}, these two are not quite equal; see the refinement in Answer 6.9. It is σ_{horiz} we are most concerned with.)

If σ_{horiz} increases in a linear way with depth, the mean value is 140 MPa, but this now operates on an area 16/13 times larger than before, or 28.8 km^2, for a total of 40×10^{10} tons; so without crushing the rock, we can now deliver enough force to keep it moving; see Figure 6.14. Of course this conclusion depends on the existence of the weak bedding planes and would be quite wrong without them.

Wedge-Shaped Thrust Sheets

The relationships among stresses become particularly simple and interesting if the sheet considered tapers right down to a point. A wedge with horizontal base is discussed first and a wedge with inclined base is treated later.

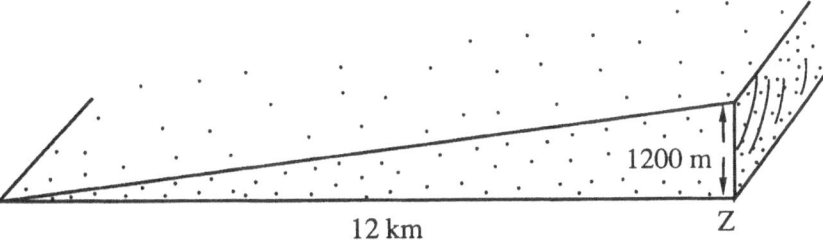

FIGURE 6.15

Question 6.10 Sliding wedge with horizontal base:

(a) Consider the wedge in Figure 6.15, whose thickness is everywhere 1/10 × distance from tip; use the lowest envelop from Figure 6.11, where shear stress = 0.4 × normal stress present, and assume vertical stress due to overburden = 30 MPa/km depth. What is the average shear stress needed at the base to keep movement going?

What is the average compressive stress needed at the back of the wedge to keep movement going?

If this average value (call it X) is provided by a set of stresses that increase from 0 at the surface to $2X$ at point Z, what compressive stress is needed at Z? How does the horizontal compressive stress needed at Z compare with the vertical overburden stress at Z?

(b) Repeat for a steeper wedge, where thickness = 1/5 × distance from tip and the bedding plane at the base is more slippery (shear stress needed for slip = 0.2 × normal stress present).

The conclusion in Answer 6.10(b) is a general result: if the thickness factor for the wedge (e.g., 1/5) equals the friction factor for the basal plane, the wedge's own lithostatic pressure is enough to make it move. For a given friction factor, wedges steeper than this will collapse under their own weight, or would except for two points that have been brushed over.

The first point is illustrated in Figure 6.16. If there is a shear stress of 14.4 MPa at Z on a horizontal plane, there must be one also of 14.4 MPa at Z on a vertical plane; but farther along to the left by a distance, say one-tenth of the way to the tip, the shear stress magnitudes will be less, about 13.0 MPa. These shear stresses on vertical planes will presumably diminish toward the surface, but all the same, they affect the vertical force balance for the segment illustrated: the force downward on its base is not exactly the segment's weight.

The second point is that we have assumed the basal plane, with shear strength 1/5 × normal stress, is the weakest plane available for slip; that is, the rock in general has more shear strength than this. The horizontal stress then need not be *equal* to the overburden stress: it could be more, or less, according to the local history and current processes.

In summary, the result reached is an approximate working rule; it is very useful, but its inherent errors need to be kept in mind. It works best when the basal plane is truly

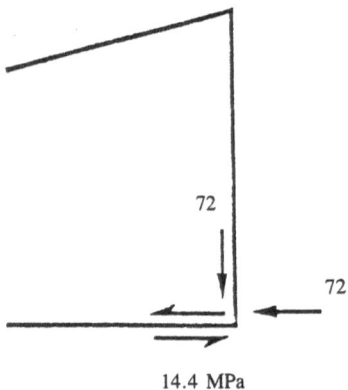

14.4 MPa

a

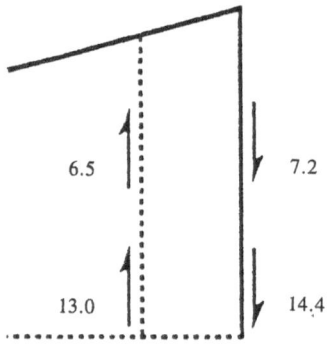

b

FIGURE 6.16

weak and the main body of rock is shot through with shear planes so as to be, in general, not much stronger. At depths less than 1 km, the lithostatic assumption becomes very hard to justify, and at depths greater than 10 km, truly weak planes are hard to imagine; but for wedges a few kilometers thick, the rule can be useful, given a suitable basal plane.

As a refinement, the effect of the vertical shear stresses shown in Figure 6.16(b) can be estimated and allowed for.

Question 6.11 Forces due to shear stress on vertical planes: Consider a portion of the wedge that stretches from the back end for a distance of 1.2 km toward the tip, which is one-tenth of the distance to the tip as already discussed. The height is 2.4 km at the back and $9/10 \times 2.4 = 2.16$ km at the front; consider a length of L km perpendicular to the page.

(a) What is the vertical force on the rear vertical surface with height 2.4 km?

(b) What is the vertical force on the front vertical surface with height 2.16 km? (In place of the approximate value 13.0 MPa, use the more exact value 12.96 MPa.)

(c) What is the net vertical force in MPa-km²?

(d) What is the net vertical force as a fraction of the segment's weight?

The exercise just completed allows us to make a small correction. If the downward force on the wedge segment's base is in fact larger by 1/25 than what we first supposed, the shear stress needed for slip will also be larger by 1/25: its maximum value will not be 14.4 MPa but closer to 15.0 MPa. Perhaps more important, the exercise brings us to the following conclusion: concentrating our attention on horizontal and vertical planes is an effective way of proceeding rapidly to an *approximate* analysis of the situation, but it is a clumsy way of seeking an exact solution. As an alternative, we can consider planes at other inclinations, using a Mohr circle if necessary to keep track of stress magnitudes. In the special case of a wedge that tapers to a point, this approach is effective, as follows.

Wedge with Inclined Base   An attribute of wedges that was noticed in the preceding section but not fully exploited is their proportionality. As Figure 6.17 shows, *if* driving stresses increase in a linear way downward from the surface, and *if* resisting shear stresses increase in a linear way inward from the tip, then *if* set (i) just balances set (ii), we can expect that set (iii) will balance set (iv). In short, if the wedge can move anywhere, it can move everywhere.

The simplicity of this setup does not depend on the slip surface being horizontal; it depends only on the stress magnitudes increasing in a linear way with distance from the

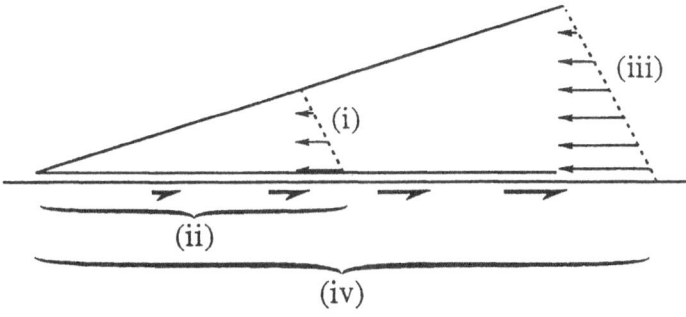

FIGURE 6.17

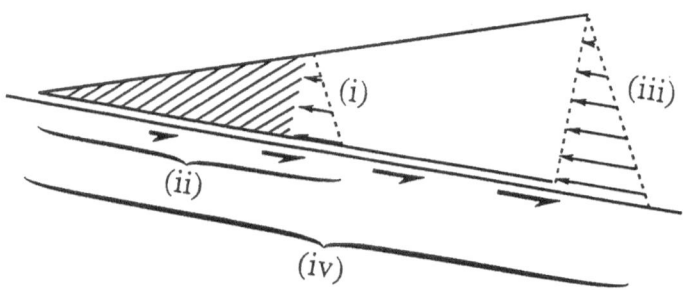

FIGURE 6.18

wedge tip, or with depth beneath the sloping surface. In Figure 6.18, the weight of the shaded rock mass not only creates the frictional resistive stresses, set (ii), the weight also has a component parallel to the slip surface of $W \sin \alpha$. But it remains true that *if* the driving stresses, set (i), can overcome set (ii) plus $W \sin \alpha$, then driving set (iii) will be able to overcome set (iv) plus $4W \sin \alpha$; or more generally if we switch from the first triangle to a triangle whose dimensions are all larger by a factor F, all forces acting on the triangle increase by a factor F^2, and remain in balance.

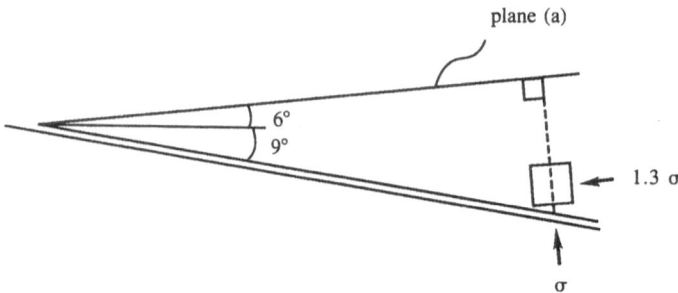

FIGURE 6.19

Question 6.12 Sliding wedge with inclined base: A rock of density 3000 kg/m² forms a topographic slope of 6° up to the east; see Figure 6.19. A weak plane in the rock slopes at 9° down to the east. The compressive stress parallel to the topographic surface is everywhere 30 percent greater than the compressive stress normal to the surface.

(a) What coefficient of friction on the weak plane would permit the wedge to slide up the slope? (Use Answers 3.34 and 4.9.)

(b) Draw a diagram like Figure 6.19 but larger, so as to show on it *three* small representative squares: one as shown with an edge parallel to plane (a), one with an edge parallel to the weak plane, and one with an edge parallel to the direction of the maximum compression. Write in the magnitudes of the stresses on all these squares, so as to see that the values make sense.

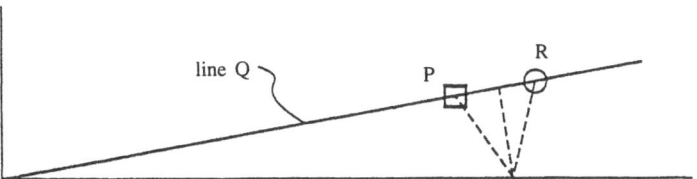

FIGURE 6.20

Question 6.13 Sliding wedge with horizontal base, continued: Consider again the wedge from Questions 6.10(b) and 6.11, with horizontal base and surface slope 1:5 (1 km vertical per 5 km horizontal distance, $\alpha = 11.3°$); we wish to check the approximate magnitudes in Figure 6.16.

(a) Use the method of Answer 3.34 to find the normal and shear stresses 2.4 km vertically beneath the surface on a plane parallel to the surface. The density of the rock is 3000 kg/m³.
(b) Use the stress magnitudes from part (a) to plot a point such as P in Figure 6.20. The line Q

shows the failure condition for the weak horizontal surface, with friction coefficient 1/5. We seek stresses that are represented by a point on this line; also, since the horizontal plane makes 11.3° with the plane represented at P, we seek stresses represented by a point 22.6° away from P around a circle whose center is on the axis. Locate a point R that satisfies this condition, and read off the normal and shear stress magnitudes that go with point R.

(c) (*Optional*) Satisfy yourself that if Figure 6.20 is correctly drawn, the radius to point R is vertical.

The conclusion from Answers 6.12 and 6.13 is that if stresses in a wedge increase linearly as in Figure 6.18, and slip behavior is governed by a fixed friction coefficient, the analysis is reasonably simple: for an internal surface parallel to the topographic slope

$$\text{normal stress} = hD \cos \alpha = dD \cos^2 \alpha$$
$$\text{shear stress} = hD \sin \alpha = dD \sin \alpha \cos \alpha$$

as in the final section of Chapter 3, and if we have independent knowledge of the compressive stress parallel to the slope, Mohr circle constructions can reveal any other detail needed. Recognizing the attractive simplicity of this picture, we should briefly consider the question, How likely is it to be an acceptable approximation to the true state of stress in the earth?

The question has already been touched on in the final section of Chapter 3, and it is helpful to review ideas from that section, in particular Figure 3.34, from which part (c) is copied here as Figure 6.21. Certainly the weight of overburden increases with depth, and in a rock of uniform density the increase is linear, as shown by the circles' diameters. The shear stress on planes parallel to the slope must then increase linearly with depth as long as all stresses down column X are replicas of the stresses down column Y. As discussed in Chapter 3, this condition is not likely to be satisfied exactly but is likely to be satisfied approximately; the error here is no greater than the error made in assigning uniform density to the rock. The greatest uncertainty attaches to the idea that compression parallel to the slope increases linearly with depth. In general there is no reason to suppose it will do so, unless the rock is currently behaving like a deforming pile of dry sand.

The characteristic of a pile of dry sand is that its failure envelope is a straight line through the origin, as in Figure 6.22. If dry sand is piled between two vertical walls and

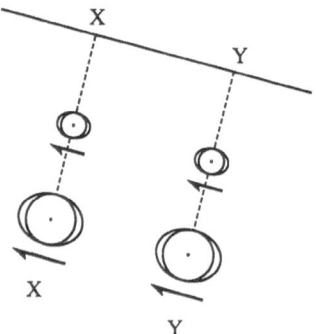

FIGURE 6.21

the walls are then moved together, the sand has to deform at all depths simultaneously. At shallow depth, the grains are under little compression σ_v and hence a small horizontal stress σ_h produces motion, whereas deeper, where σ_v is larger, σ_h has to build up to a larger magnitude: the sand grains shuffle about until this state is reached. In other words, if a pile of dry sand is deforming everywhere, then as long as deformation continues, the horizontal compression *has to be* proportional to depth; and if the rock beneath a sloping topography behaves similarly, the same linear increase of compressive stress must be present.

The behavior of dry sand can be tested in a benchtop experiment and is reasonably well in accord with the foregoing ideas: a small wedge can be fed with sand so that it

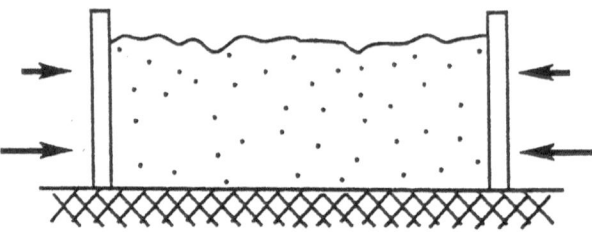

a

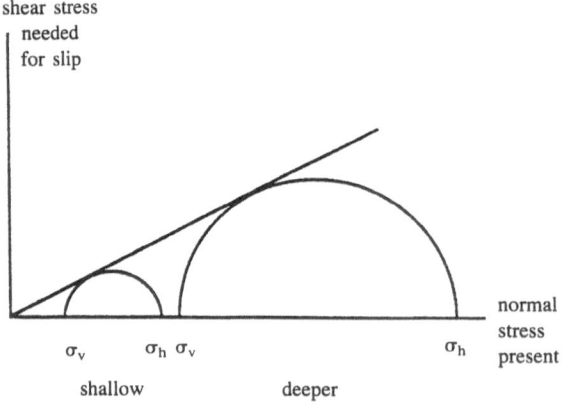

FIGURE 6.22

FIGURE 6.23

grows larger, and the later large wedge is approximately a scaled-up copy of the early small wedge (see Figure 6.23). But the behavior of rock in the earth's crust is considerably affected by the presence of fluids; it is necessary to progress into the next section before the usefulness of the dry-sand analogy can be properly judged.

EFFECT OF PORE FLUID ON PARTING

As soon as attention is given to a rock with pore fluid, the possibility arises of defining and using an **Effective Stress.** Individual preference varies in this matter: some people favor the concept and use it as widely as possible, while others use it in a narrower range of circumstances. In the present text, an attempt is made to accommodate this range, as follows: ideas are introduced without effective stress being mentioned, but in a subsequent section the same material is covered using that concept. It is hoped that this will leave readers free to decide for themselves how far effective stress is a convenient short cut, and how far it is an oversimplification by which we lose track of effects that we would prefer to keep clearly in view. In short, the intention is to be nondogmatic, at some sacrifice of brevity.

Tensile Parting

Experiments of the type shown in Figure 6.24 (made famous by P. W. Bridgman) showed that parting in the tensile manner is just as likely in all four; the material fails when

$$\text{tensile stress } + \text{ fluid pressure } = \text{ tensile strength.} \tag{6.1}$$

The internal stress distribution on a scale of grains is of some interest (see Questions

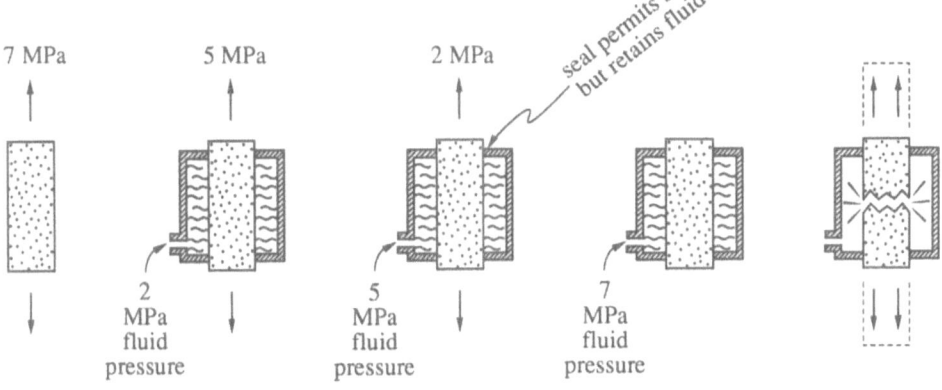

FIGURE 6.24

3.18 and 3.23) but, without inspecting those details, we can look right away at geological consequences.

Hydrofracture and Vein Formation   A fifth experiment in the series from Figure 6.24 can be imagined, where a mild *compression* is exerted on the ends of the sample. It is still, presumably, possible to part the sample in the tensile manner if we reach the condition

$$\text{fluid pressure} = \text{tensile strength} + \text{compressive stress.} \qquad (6.2)$$

This equation underlies the practice of hydrofracture ("fracking") that is widespread in the oil business. A portion of drillhole is sealed with packers above and below the interval to be fracked, and fluid is pumped into the space between the packers. At a sufficient pressure, the rock formation around the fluid pocket is split open by the fluid pressure. The increase in permeability makes the well more productive (after the fracking equipment has all been cleared away).

It is reasonable to suppose that veins are formed by a similar process, and also simple joints (with no mineral filling or separation of the joint walls). There is *some* fluid quantity and fluid pressure at every grain interface in the earth, and always the possibility that Equation (6.2) will be satisfied. Three points can be noted:

1. Partings may become filled or remain unfilled. In hydrofracturing, the fluid used is sometimes loaded with sand or other solids that lodge in the fractures created and prevent their closing up again; thus artificial fractures are of two kinds, filled and unfilled—and joints and veins are a similar pair. A joint is a fluid-created parting that has not acquired a filling. In addition, many "solid rocks" are probably rich in surfaces that have at times parted, for reasons of fluid pressure, and have subsequently not just closed but healed as well.

2. The *tension* aspect is perhaps unnecessary and misleading. We could rewrite Equation (6.2) to state that an intrusion occurs when the intrusion pressure exceeds the confining pressure by a certain amount, that could be called the entry stress or entry threshold; thus

$$\text{intrusion pressure} = \text{confining pressure} + E. \qquad (6.3)$$

This terminology would directly match the geological circumstances, where the confining pressure is great and the compression that forces the intrusion in is even greater. There may well be local tensile stresses in a tiny region around an advancing crack tip, that resemble local tensile stresses in a genuine tension test, but attention to these is optional, not necessary; the terms we *need,* in order to write out mechanics equations for outcrop observables, are the terms in Equation (6.3).

3. Points 1 and 2 taken together bring up the thought that it is not only water that can be forced into rocks, opening them up. Any mobile material subjected to pressure at a high-pressure site will tend to create cracks in a neighboring material at lower pressure. Obviously we have not only hydrofractures but magma fractures (i.e., dikes) and natural-gasofractures. But other constituents of the intergranular mess also exert pressure; if squeezing a sandstone from north and south forces quartz to disappear from east-west interfaces and to force itself into north-south interfaces, it begins to create quartzofractures (Figure 6.25). Even here, water catalyzes the migration but quartz (or calcite or any active constituent) can realistically be considered a pressure-transmitting mobile material.

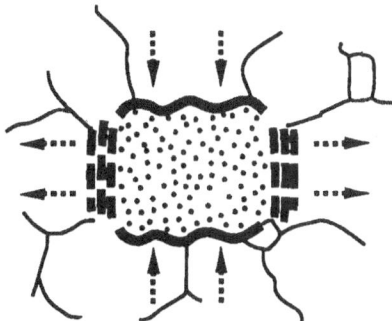

FIGURE 6.25

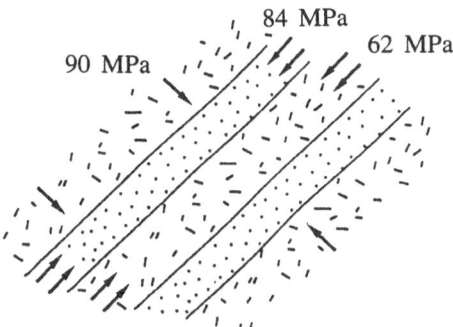

FIGURE 6.26

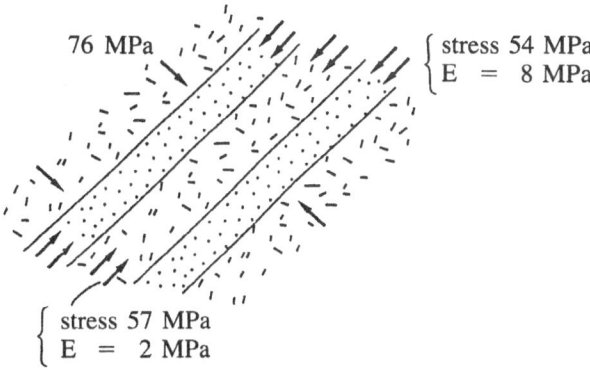

FIGURE 6.27

Question 6.14 Boudinage: The topic is continued from Question 5.15.

(a) If both rock types in Figure 6.26 have entry pressure $E = 3$ MPa, at what fluid pressure will boudinage occur? Which layer will part?

(b) Repeat for the values of E and stress states shown in Figure 6.27

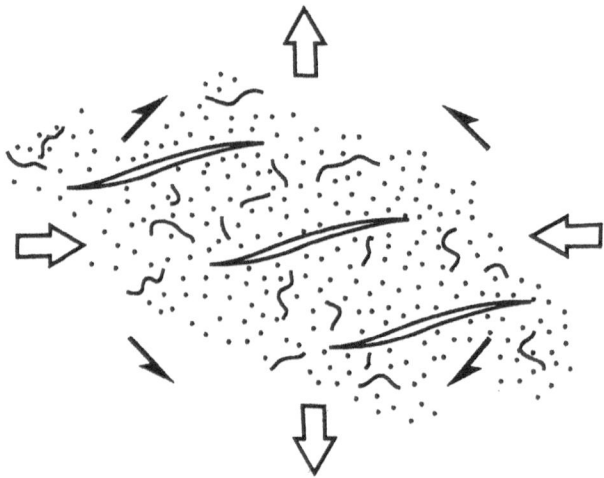

FIGURE 6.28

Question 6.15 *En echelon* vein sets: If E for the rock shown in Figure 6.28 is 2 MPa, mean stress = 21 MPa (at 700 to 900 m depth) and maximum shear stress = 5 MPa, what fluid pressure will cause the veins shown to form? What is the ratio fluid pressure/mean rock pressure?

Question 6.16 *En echelon* vein sets, continued: If E for the rock is 2 MPa, mean stress = 21 MPa, fluid pressure is 18 MPa and rock viscosity is 3 gvu, at what shear strain rate will *en echelon* veins appear?

A point of interest is that as soon as a set of veins opens, the effective viscosity of the zone containing them goes down, perhaps by as much as a factor of 10. This is, then, an example of an idea that has been mentioned before (discussion of Answer 5.17) in a nonspecific way: an increase of stress can alter the number and arrangement of weaknesses in a rock in such a way that the rock as a whole becomes weaker—as stress goes up, effective viscosity goes down. The effect is discussed again in Chapter 7 under the heading A Changing Population of Microfractures.

Note: If one totally ignores the development of fractures and uses instead a description of rock behavior that states

$$\text{strain rate} = k \cdot (\text{shear stress})^n,$$

then if $n > 1$, as stress goes up, effective viscosity goes down. Thus for a series of observations of strain rate and stress, values of k and n can be found that make the formula match the observations. But the equation is something of a trap, for the following reason. The shear stress is the outside agent that acts on the rock, and this seems to leave k as the rock property that controls the magnitude of the rock's response. But as just noted, when stress increases the rock's texture changes and the rock becomes intrinsically more mobile: the rock's mobility is itself a function of stress. The equation could in fact be written

$$\text{strain rate} = \underset{\text{rock mobility}}{[k(\text{stress})^p]} \cdot \underset{\text{driving agent}}{(\text{stress})^q}$$

with $p + q = n$. To write the equation in this way helps to keep the actual rock processes from being lost to view, even though we have more or less to guess how n is partitioned into p and q.

Shear Parting

If one looks at the present disposition of geological materials around the earth, one might say that it is affected more by shear parting than by any other process. Besides the mighty strike-slip faults—Altyn Tagh, San Andreas, New Zealand's Alpine Fault—every subduction zone is a shear parting, and every mountain range is built at least partly out of slabs emplaced by shearing or slip with respect to each other. Of course, at mid-ocean ridges, tensile parting has also been responsible for thousands of kilometers of motion; the section on tensile parting has considerable relevance to global geology, as well as the present section. Nonetheless shear parting is one of the earth's essential processes from the smallest to the largest scales; and at every scale, as far as one can tell, it is pore-fluid activity that catalyzes the action.

The magnitude of a pore fluid's effect can be shown in diagrams such as Figure 6.29 and 6.30 which are expansions of Figure 6.11. Consider first intact rock, Figure 6.29. With no fluid pressure effect, one of the conditions at which the rock fails is

$$\sigma_{min} = 20 \qquad \sigma_{mean} = 80 \text{ MPa}$$
$$\sigma_{max} = 140 \qquad \tau_{max} = 60 \text{ MPa}$$

Question 6.17 Effect of fluid pressure on shear failure:

(a) For σ_{min} still = 20 MPa as before but fluid pressure 20 MPa, at what magnitude of σ_{max} does the rock fail? What is τ_{max} at the moment of failure?

(b) For σ_{max} = 140 MPa as before but fluid pressure 20 MPa, at what value of σ_{min} does the rock fail? What is τ_{max} at the moment of failure?

(c) For σ_{mean} = 80 MPa, how much *difference* $\sigma_{max} - \sigma_{min}$ can the rock withstand
    (i) for fluid pressure = 20 MPa
    (ii) for fluid pressure = 40 MPa?

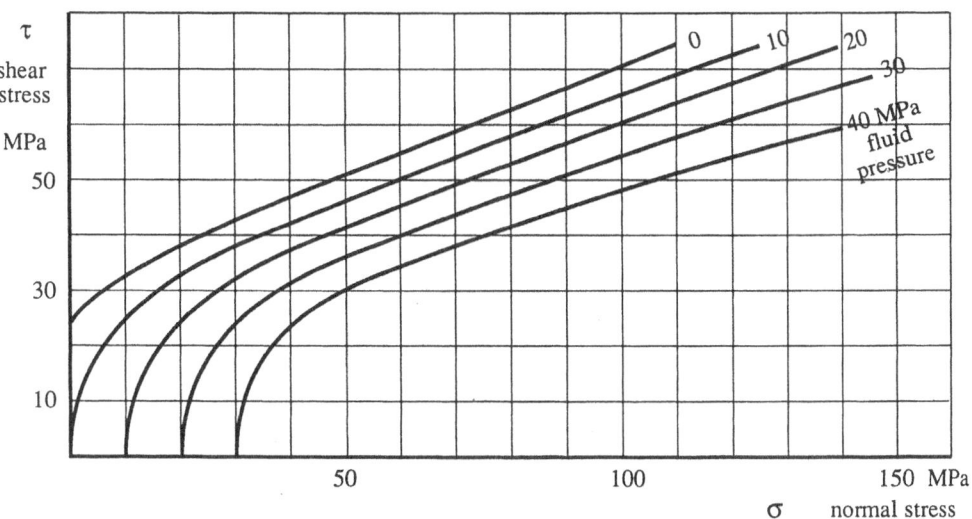

FIGURE 6.29

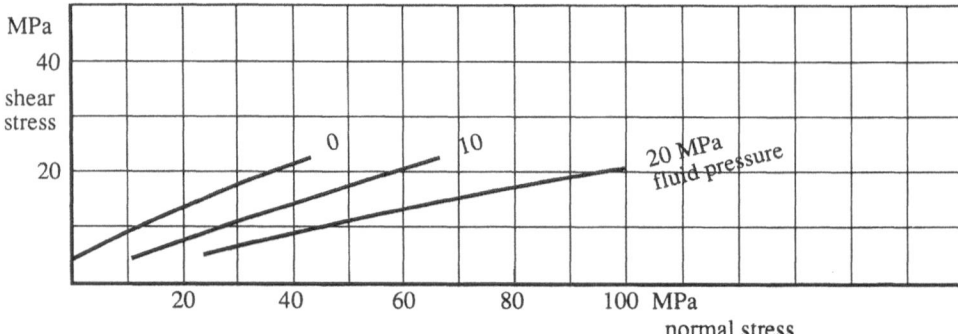

FIGURE 6.30

The conclusion from Question and Answer 6.17 is that fluid pressure makes fracturing easier or more probable. Simply looking at Figure 6.29 in fact shows this effect without any numbers: at a steady value of normal stress, the higher the fluid pressure, the lower the shear stress the rock can withstand; or at a steady value of shear stress, the higher the fluid pressure, the higher the normal-stress magnitude needed to keep the rock intact.

How then might the presence of a fluid affect the inquiry in Question 6.9? The outcome of that inquiry was that a gradually increasing force acting on one side of the slab would fracture the slab before causing it to slide. In presence of a fluid, the slab would fracture more readily, as just noted. But it would also slide more readily, so that to complete the inquiry we need to look at Figure 6.30.

Question 6.18 Effect of fluid pressure on thrusting:

(a) For a vertical compressive stress of 39 MPa, what shear stress on the horizontal basal plane is needed to cause slip if the fluid pressure is 10 MPa?

(b) How much normal stress applied uniformly all over the slab's long edge would generate enough force to cause slip?

(c) Can you devise a nonuniform distribution of stress that has a mean value as great as the value in part (b) and does not cause failure in intact rock at any level
   (i) if fluid pressure = 10 MPa throughout the slab?
   (ii) if fluid pressure diminishes linearly from 10 MPa at the base to zero at the surface?

The conclusion from Answer 6.18 is that the sliding of a slab on a horizontal surface is facilitated by fluid pressure. A similar effect occurs with a slab on an inclined surface, as in Figure 6.12; and considering such a slab brings attention to a second noteworthy aspect of fluid behavior.

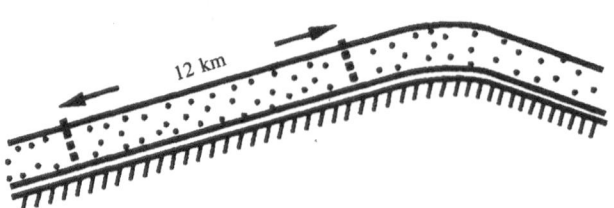

FIGURE 6.31

Question 6.19 Effect of fluid pressure on downhill sliding:

(a) Let the slab of rock in Question 6.9 rest on a slope of inclination α where $\sin \alpha = 5/13$ and $\cos \alpha = 12/13$ but let it be part of a continuous sheet that extends over a crest as in Figure 6.31. Let the behavior at the base of the slab be as in Figure 6.30 and let the fluid pressure at the base of the slab be 10 MPa. For the 12-km segment shown, estimate
   (i) the component of the slab's weight that acts parallel to the slope
  (ii) the stress normal to the slope that arises from the slab's weight
 (iii) the greatest shear stress that can act on the base without initiating slip.

(b) Multiply your answer at (a.iii) by the area of the base to calculate the greatest shear force that can act parallel to the base. This force is less than your answer at (a.i) and we have to consider whether enough other forces are present to stabilize the slab. Other effects resisting motion are a compressive stress at the lower end and a tensile stress at the upper end. Let us suppose that these are in the ratio 3/4:1/4. How does the average tensile stress at the upper end compare with the tensile strength of the intact rock? How does the sum of (average tensile stress) + (fluid pressure) compare with the tensile strength of the intact rock? Take tensile strength = 10 MPa as in Figure 6.11.

The conclusion from Answer 6.19 is that the tensile stress by itself is too small to part the rock, but that the combination (tensile stress) + (fluid pressure) exceeds the tensile strength, so that the rock layer is likely to part at the upper end *and let the fluid escape.* This is an important general feature of fluid behavior. Suppose the fluid pressure at the base of the slab is initially low but is increasing. Initially the shear stress at the base is large enough to support the slab. As fluid pressure rises, the maximum shear stress possible at the base drops down and the slab comes to depend for stability on the compressive and tensile effects at its ends. Further rise of fluid pressure increases *both terms* on the left of Equation (6.1).

tensile stress + fluid pressure = tensile strength

until tensile parting occurs. But as already noted, parting provides an easy channel or channels for fluid to escape, and hence the pressure beneath the slab is likely to diminish again. This is just one instance of a common phenomenon: increase of fluid pressure very often creates improved channelways—new partings, or a widening of existing channels, or an increase in intergranular permeability; quite commonly the system "overshoots" and creates more channels than just enough to cope with the original influx; there is then an excessive escape of fluid, and fluid pressure drops back down and allows the channels to close up. (In the configuration of Figure 6.31, there is no reason for the upper-end tensile partings ever to close, but most hydrofractures form deeper in the earth and positively tend to close and heal when fluid pressure drops.) Thus the overall behavior is spasmodic. Everyone knows that volcanoes are spasmodic, and probably a great many more fluid movements through the earth are spasmodic as well; we have just seen reasons for expecting them to be so.

A Rock's Response Affects Its Properties   It has just been emphasized that entry of a fluid affects a rock's permeability, i.e., affects the ease of entry. There is a parallel with the point made in Chapter 5, that when a rock deforms in a geometrically continuous way, shearing the rock affects the ease of shearing. These can be thought of as two *separate* instances of self-accentuating behavior. But there is a third truth, a cross-link between the first two: shearing a rock affects the ease of entry of fluids (Questions 6.15 and 6.16 show an example).

The reason behind the assertion just made concerns microfractures. The rock's stiffness or viscosity, its ability to respond slowly to shear stresses, is affected by its population of microfractures, and the population is in turn affected by the rock's changing shape. But the microfractures are the fluid's channelways: the change of shape that we treat as geometrically continuous in fact involves micropartings that admit fluid that stimulates macropartings. And similarly on the scale of a mountain belt: an outcrop where veins are forming, or fractures are opening, is a tiny element in the larger system, whose bulk response enters into the mechanics of crustal shortening, mountain uplift, folding, and thrusting. What is seen as an individual fracture on one scale is summarized in a continuum description on another scale. Thus again, fluids in fractures affect processes that we describe in terms of a geometrically continuous deformation. In this chapter and the preceding one, geometrically continuous behavior and fracturing have been treated as separate topics, but there are severe limits on what can be achieved by that approach; geomechanics comes fully to life when the two behaviors are meshed in a single treatment, as in Chapter 7.

Effective Stress

People who try to understand the structures seen in rock outcrops and hillsides carry two ideas in mind. On one hand we picture a rock as a continuum, in which the stress state, strain rate, and so on vary continuously from point to point; on the other hand we picture a rock (more realistically) as an aggregate of grains that is shot through with grain boundaries, fractures, and other kinds of flaw that are filled by a *fluid*. The stresses that grains in a rock exert on one another can vary greatly over a few millimeters from point to point and from one direction to another, whereas one can suppose that the fluid in the "pore space" is not so variable; the pressure it exerts is more uniform from one direction to another as well as being more uniform from point to point. These two pictures, of the continuum and of the pores, have already been used in constructing Figure 6.29: that diagram is based on the idea that for a plane inside a material of interest, we can identify a normal stress, a shear stress, and a single fluid pressure. Assuming that these three quantities can be measured or imagined, we then define

effective normal stress = (normal stress) − (fluid pressure),

or, as more commonly written,

effective stress = total stress − fluid pressure.

(As regards shear stress, no adjustment for the presence of a pore fluid is made.) The purpose of the definition is to make a particular group of ideas more compact. Like most simplifying assumptions, on one hand it is truly useful within its proper field of application; on the other hand it can be an obstacle when we try to consider phenomena that lie outside its field.

Tensile Parting   Refer to Figure 6.24: to the extent that the arithmetic relation suggested in the diagrams is borne out by actual tests, we can say the rock parts when the effective stress = −T; this statement is exactly equivalent to Equations (6.1) and (6.2).

Shear Fracture   Refer to Figure 6.29: to the extent that all the curves shown have the same shape, Figure 6.29 can be replaced by Figure 6.32. In most series of laboratory tests, the errors introduced by replacing separate curves by a single average curve are no larger than other effects inherent in the test program, such as variability from sample to

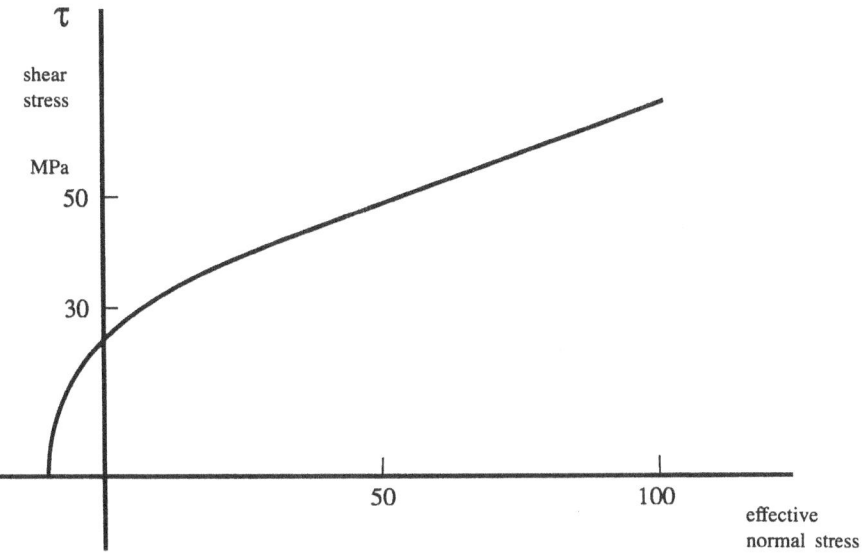

FIGURE 6.32

sample of the parent rock being reported on. But this condensation perhaps leads us to neglect an effect that may be important on the geological time scale, as discussed in the next section.

Question 6.20 Effect of fluid pressure on shear failure, continued: Repeat Question 6.17 using Figure 6.32 instead of Figure 6.29.

Effective Coefficient of Friction   After a dry material has failed, if slip is occurring on a single clearly identifiable fracture plane, we expect that within reasonable limits of error

shear stress needed for slip = μ (normal stress on fracture plane),

where μ is a coefficient of friction that can be treated as constant over a wide range of normal stresses. But if the material carries a pore fluid, the shear stress needed for slip will be affected by the fluid pressure as well as by the normal stress. In ideal circumstances we would find

shear stress = μ (normal stress − fluid pressure),
needed for slip

and the right-hand side can be contracted in two different ways. We can write either

shear stress = μ (effective stress)
needed for slip

or

shear stress = μ' (normal stress).
needed for slip

Here μ' is an effective coefficient of friction, defined as μ (normal stress − fluid pressure)/normal stress. Either way, if for example the fluid pressure is 0.6 × normal stress, the shear stress needed for slip is reduced to 0.4 of its magnitude for dry rock. In either version, we are describing only an idealization, but the two versions together help us to keep in mind an idea that is probably needed for geological slip: a fluid acts *both mechanically* in carrying part of the compressive load *and chemically* in reducing the hindrance caused by surface roughnesses.

Question 6.21 Effective coefficient of friction:

(a) In discussing thrust sheets, we used coefficients of friction of 0.4, 0.2, and 0.15. If the surface discussed had dry coefficient of 0.8, what value of the ratio (fluid pressure)/(total normal stress) would give an effective coefficient of 0.4? of 0.2? of 0.15?

(b) In a rock where the dry coefficient is 0.6, the fluid pressure has an average value of 0.8 × (total normal stress) but varies locally from 10 percent less to 10 percent more than this. What are the effective coefficients for the least fluid pressure, the average and the maximum fluid pressure? Using the middle one of your three answers as reference value, by how much percent are the other values smaller or larger?

The purpose of Question 6.21(b) is to emphasize the large variation that shows up in the effective coefficients. The coefficient is controlled by the *difference* between fluid pressure and total pressure, and when this difference becomes small, its fluctuations become large, in relative terms. A second point is that slip surfaces are like chains: the *average* strength is of very little interest—it is the strength of the *weakest part* that is of most concern. This means that, regarding fluid pressure, again the average value is not of as much interest as the maximum value it reaches locally; wherever the fluid pressure is locally high, that is the site where the rock will fail.

The exercise just completed emphasizes the mechanical role of the fluid, i.e., its role in reducing the effective stress. To avoid the error of taking an overly simple view, we should consider again the character of real fractures or fault zones, which are both variable and complicated. In feldspathic and pelitic rocks, they can accumulate clay, and in dolomitic and ultramafic rocks they can accumulate talc, whereas in quartzites they mostly accumulate fine-grained quartz. In any rock, the slip surface accumulates rock fragments that sometimes help by acting as rollers and sometimes hinder by catching on projections. Regardless of the variability, it seems clear that rearrangement of material by dissolution and reprecipitation can be more important on a geological time scale than in a laboratory experiment: given time in which to act, a pore fluid can reduce the coefficient of friction by being chemically active *as well as* reducing the effective compressive stress by being mechanically active. It is for this reason that in Figure 6.30 the failure envelope at high fluid pressure has been given a lower position and flatter slope, and is not simply the envelope from low fluid pressure translated sideways. A similar effect has also been introduced in Figure 6.29 but not to such a noticeable extent.

Summary The effective stress is defined as the difference between pore-fluid pressure and the total normal stress across a planar element inside a rock. In conditions where the rock behavior is straightforward and the effect of the fluid is only mechanical (inert chemically), the rock's failure behavior can be well summarized in terms of effective stress. But using the concept of effective stress commits us to simplifying assumptions—to assuming that fluid pressure and total stress boil down to just a *single* variable, whereas geological realities sometimes suggest otherwise. The following supposition cannot be verified by laboratory experiments but all the same, it is supposed that on geological time scales, rock failure is not well described by effective stress, and that to use two separate variables as in Figures 6.29 and 6.30 is more satisfactory.

ANSWERS

Answer 6.1

(a) 13.0 MPa   6.7 MPa   4.8 MPa   5.0 MPa
    8.0 MPa
(b) 1.0 MPa   3.5 MPa   3.8 MPa
(c) The acute angle between intersecting fractures be-
    comes more acute as you go to lower pressures in
    the common case where the failure envelope is
concave downward. (We refer here to the angle
between the fractures when first formed. The angle between two fractures as seen in an outcrop is
not necessarily the angle at which they first
formed.)

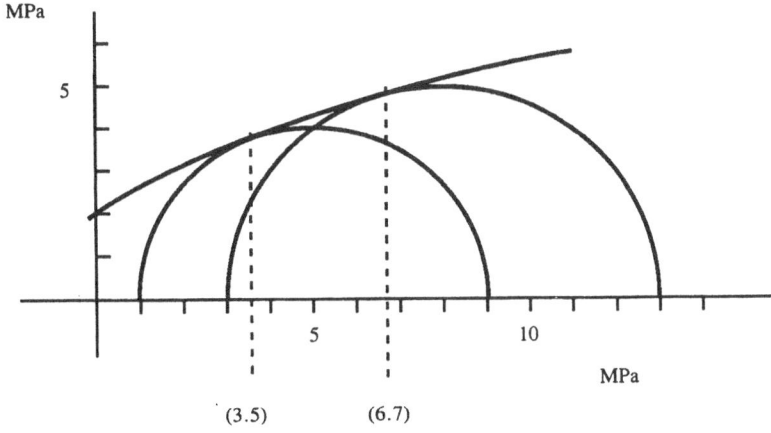

FIGURE 6.33

Answer 6.2

(a) A stress difference of 22 MPa (giving σ_{max} =
    31 MPa and σ_{min} = 9 MPa) might serve. The
    envelope one inserts on the basis of the data can be
    straight or curved to some degree, and so there is
    no unique answer that is definitely correct.
(b) Again the details depend on exactly how an enve-
    lope is drawn to fit the data. For the envelope
    shown in Figure 6.34, the rock fails in the shearing
    manner first.

(c) According to the idealized diagram, the rock
    would fail first in the tensile manner. But real
    rocks are full of irregularities and the stress state is
    likely to vary from point to point; one should ex-
    pect that a series of similar samples in test B or test
    C would give rather varied responses, failing part-
    ly in tension and partly in the shearing manner. It
    is fruitless to attempt firm predictions in cases
    such as these where the stress circle and the enve-
    lope have similar curvatures.

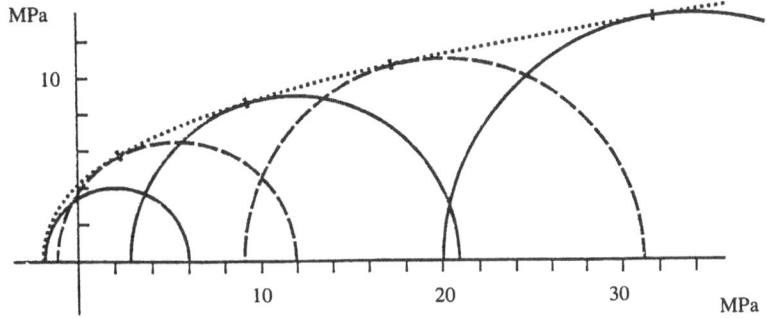

FIGURE 6.34

Answer 6.3

Plane *g* is less likely to fail than plane *e* because the extra normal compression helps to keep the grains together; the greater the normal compression, the more the *roughness* of a possible slip plane hinders it from slipping.

Another possible answer, "Plane *e* is more likely to fail because the failure envelope slopes upward . . . ," contains the action-at-a-distance mistake. The envelope slopes upward because plane *e* fails first, not the other way round. The diagram does not cause the behavior, it simply records it.

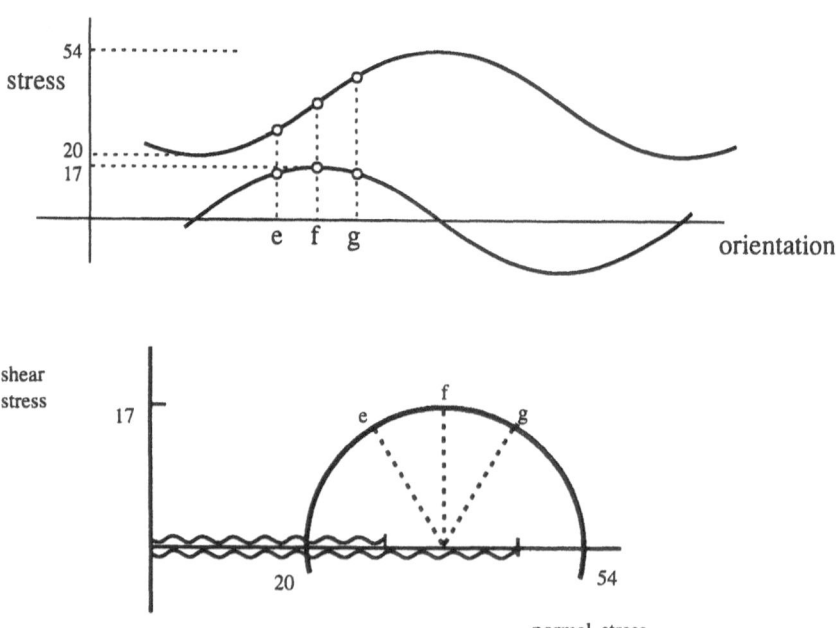

FIGURE 6.35

Answer 6.4

At 15 km depth, the overburden exerts a pressure of about 450 MPa so we need a circle of diameter 30 MPa at about 450 MPa on the normal stress axis.

A tensile strength of 3 MPa is almost too small to show.

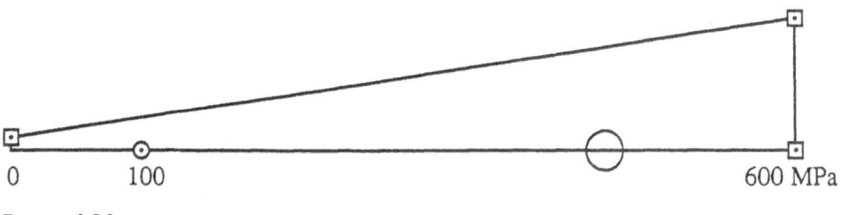

FIGURE 6.36

Answer 6.5

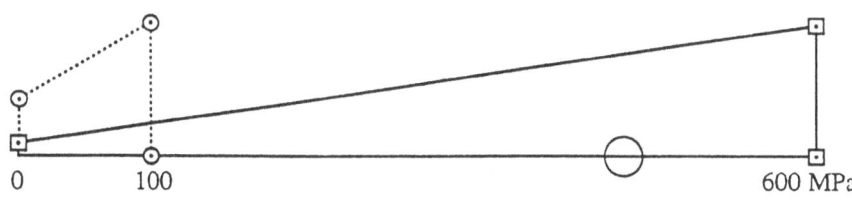

0   100   600 MPa

FIGURE 6.37

Answer 6.6

(a) If we neglect the presence of fluid in the hole, the stress state has $\sigma_{min} = \sigma_{horiz} = 0$,

$$\sigma_{max} = \sigma_{vert};$$

then, to generate a shear stress of 60 MPa, we need

$\sigma_{vert} = 120$ MPa, as found at 4000 m depth.

(b) To allow for fluid pressure and to focus on a depth of 4800 m, we calculate

σ_{vert} due to overburden = 144 MPa,
σ_{horiz} due to fluid = 60 MPa,
max shear stress = 42 MPa;

thus we would *not* expect the wall-crushing effect shown. (In fact at 6800 m, conditions are close to failure; this is about 23,000 feet, which is deeper than most drillholes. The fluid not only keeps natural brines out of the hole, as in Question 3.7, but also performs a useful function helping to stabilize the walls.)

Answer 6.7

The relevant stress circles are already drawn. For reverse faulting we need $\sigma_{max} = \sigma_{horiz} = 48$ MPa. For normal faulting $\sigma_{horiz} = \sigma_{min} = 3$ MPa. A typical depth for $\sigma_{vert} = 20$ MPa is 700 or 800 m.
The normal stress on the normal fault is about 10 MPa, while on the reverse fault, it is about 30 MPa.

In a simple world, somewhat more crushing might be expected on the reverse fault. But *caution:* most fault processes are much affected by fluids; the present discussions, with fluids ignored, are only preliminaries, and it would be a mistake to attempt much in the way of real-world conclusions while we are still working with elementary idealizations.

Answer 6.8

As in Figure 6.6, the angles of intersection of a conjugate pair of faults are the angles that appear in the Mohr diagram, in this instance 74° and 106° (Figure 6.38). Hence the dip angle = 53°.
At successively smaller magnitudes of σ_{vert}, the angles change to *less than* 74° and *greater than* 106°. In the extreme, the acute angle diminishes a lot, and eventually becomes 0° for tensile fractures. Correspon-

dingly the dip angle increases from 53° toward 90° at shallower depths. If a single fault plane extended from 800 m depth to the surface, it could be curved as shown in Figure 6.39. Some normal faults do indeed have curved profiles of this type but, as noted before, in real situations, other influences are likely to be at work in addition to the stresses just described.

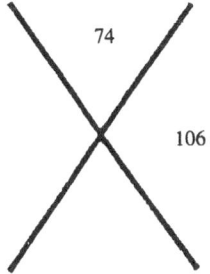

FIGURE 6.38

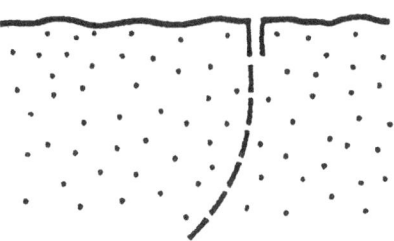

FIGURE 6.39

Answer 6.9

(a) The shear stress needed is controlled by the normal stress which, in this case, is the overburden stress or pressure, 35 or 40 MPa or in that region, according to the rock's density. For normal stress 40 MPa, the shear stresses are 21 MPa to initiate movement and 16 MPa to maintain movement once started. The total force needed then, over the relevant area of 216 km², is 45×10^{10} t to initiate movement and 35×10^{10} t to maintain it.

(b) The weight that gives normal stress 40 MPa is 86×10^{10} t, so the two forces are respectively 0.52 and 0.40 times the slab's weight.

(c) If we call the angle of tilt $\alpha°$, then $\sin \alpha = 0.52$ or 0.40, and $\alpha = 31°$ or $24°$, so the orders of magnitude are all right.

(d) The area of the slab's long edge is 23.4 km²; therefore the uniform stress that would give 35×10^{10} t is 148 MPa. From the upper curve in Figure 6.11, we find that intact rock could withstand $\sigma_{max} = 148$ MPa if σ_{min} were at least 24 MPa—a condition that is generated by the overburden for the lower part of the slab but not for the upper part.

(e) At the surface where $\sigma_{vert} = 0$, the greatest σ_{horiz} the rock can withstand is 90 MPa. Halfway down, $\sigma_{vert} = 20$ MPa and the rock can withstand 140 MPa of horizontal stress. At the base, $\sigma_{vert} = 40$ MPa and the rock can withstand 180 MPa. If horizontal pressure were applied in a nonuniform manner (less at shallow depths and greater pressure at greater depth) we could apply more force without cracking the rock than if the stress were the same at all depths; but still we cannot get the average value up as high as 148 MPa.

A refinement: at the base, $\sigma_{vert} = 40$ MPa and we seek the greatest σ_{horiz} that will just not crush the rock. Roughly, we start from 40 MPa on the normal-stress axis and draw the largest circle we can. But at the base, a horizontal surface suffers not only σ_{vert}, it also suffers the shear stress of 16 MPa (or 21 MPa to initiate movement.) Thus the horizontal surface is not a principal plane and σ_{vert} is not σ_{min} for the stress state; to find σ_{min}, strictly we should perform the construction in Figure 6.40 and seek the largest circle through the two x's that lies under the envelope, rather than the largest circle through the 40 mark on the axis. In the problem at hand, the difference is small in numerical terms; but the difference between an erroneous approach and a correct one is worth noting.

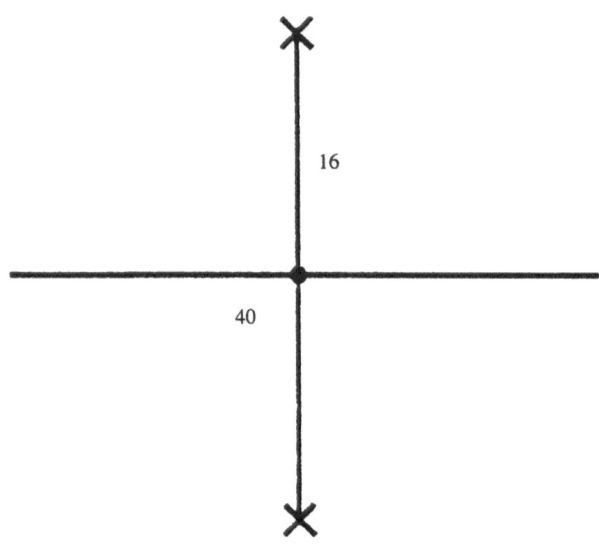

16

40

FIGURE 6.40

Answer 6.10

(a) The vertical overburden stress at Z = 36 MPa so shear stress needed at Z = 14.4 MPa.
Average shear stress needed over base = 7.2 MPa.
Average compressive stress needed at back = 72 MPa.
Compressive stress needed at Z = 144 MPa, four times as great as the overburden stress.

(b) Overburden stress at Z = 72 MPa;
shear stress needed at Z = 14.4 MPa;
average shear stress needed over base = 7.2 MPa;
average compressive stress needed at back = 36 MPa.
Compressive stress needed at Z = 72 MPa, equal to the overburden stress.

Answer 6.11

(a) The average vertical shear stress is 7.2 MPa and the area on which it operates is 2.4 L km^2 so the vertical force is 17.28 L MPa-km^2, downward.

(b) The average vertical shear stress is 6.48 MPa and the area on which it operates is 2.16 L km^2 so the vertical force is 14.00 L MPa-km^2, upward.

(c) Net vertical force is 3.28 L MPa-km^2 downward.

(d) The overburden stress at the back of the segment is 72 MPa and 9/10 × 72 MPa at the front, giving an average of 68.4 MPa on an area of 1.2 L km^2, so the segment's weight is 82.08 L MPa-km^2. The vertical force due to shear stresses, evaluated at part (c), is thus almost exactly 1/25 of the segment's weight.

Answer 6.12

(a) A helpful Mohr circle diagram has already been drawn at Answer 3.9 and forms the basis for Figure 6.41. The weak plane makes 15° with the topographic surface, plane (a); hence we can find the stresses on it at 30° from point *A* in the diagram.

Stress magnitudes are 260 MPa normal and 38 MPa shear stress; the plane needs an effective coefficient of friction of 38/260 or 0.15 (or less than this) to permit slip.

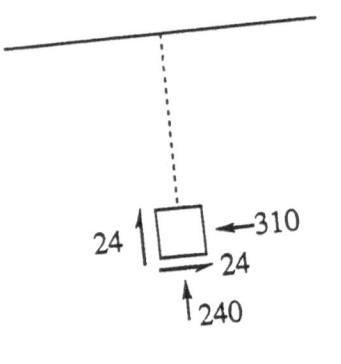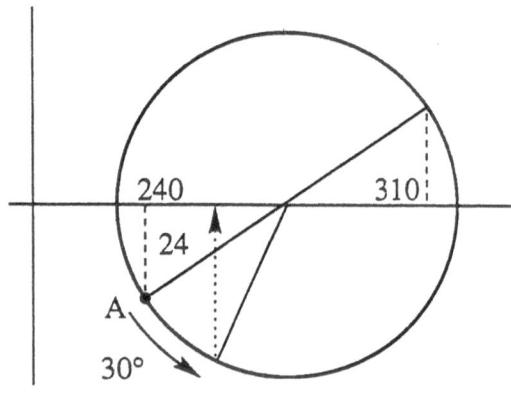

FIGURE 6.41

(b)

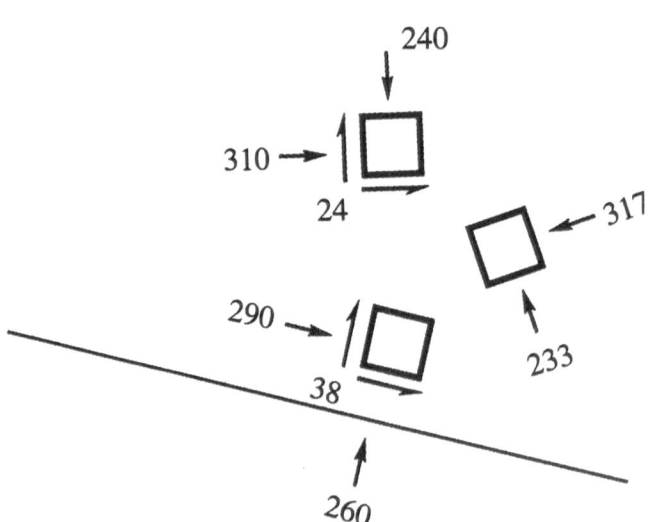

FIGURE 6.42

Answer 6.13

(a) Using the assumptions on page 80, the shear stress on a plane parallel to the surface = $dD \sin \alpha . \cos \alpha$ = (2.4 km) (3000 kg/m³) $\sin \alpha . \cos \alpha$ = 36 $\sin 2\alpha$ MPa, and the normal stress is 72 $\cos^2\alpha$ MPa. Tan α = 1/5, giving shear stress 180/13 or 13.85 MPa and normal stress 900/13 = 69.23 MPa.

(b) Normal stress 75 MPa, shear stress 15 MPa as estimated in Answer 6.11. For the point considered, at 24 km vertical depth and 12 km horizontally from the wedge tip, the complete stress state is represented by a circle passing through both P and R, with center U (see Figure 6.43). We have to remember that line Q is not a failure envelope for the whole rock; line Q relates only to the particular weak plane present. Thus plane R fails, while plane P and all other planes do not.

The radius UR is perpendicular to the axis, with 2θ equal to 90°. Also plane R is horizontal, so that the direction of $\sigma_{\max}$ must be at 45° to horizontal. This agrees with the condition shown in Figure 6.16(a): if $\sigma_{horiz} = \sigma_{vert}$, either the state is hydrostatic or the inclination of $\sigma_{\max}$ is 45.°

(c) If in Figure 6.43 we make angle (i) equal to α, line S makes 90° with line Q. Also, if line T bisects the angle 2α, line T makes 90° with line Q; thus T is parallel to S. But angle (ii) equals α, and so angle (ii) equals angle (i) and the radius to R is vertical.

For stress magnitudes, we note that if tan α = 1/5, then cos 2α = 12/13, = PV/PU = PV/RU. Thus the two similar triangles OPV and ORU are in the ratio 12/13, which confirms the stress magnitudes obtained in parts (a) and (b).

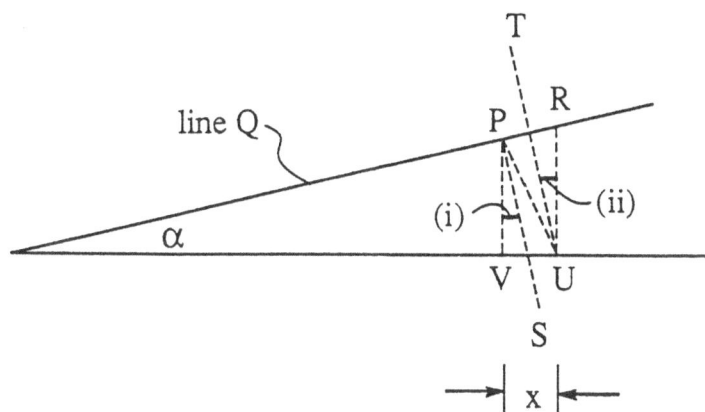

FIGURE 6.43

Answer 6.14

(a) 65 MPa; the thicker layers part first.
(b) 59 MPa, same.

Answer 6.15

From the data, $\sigma_{\max}$ = 26 MPa and $\sigma_{\min}$ = 16 MPa. Hence, the fluid pressure needed is 18 MPa, which is 6/7 or 0.86 of mean rock pressure.

Answer 6.16

Veins will appear when $\sigma_{\min}$ drops to 16 MPa, when the maximum shear stress = 5 MPa. This shear stress goes with a shear strain rate of 5/3 per gtu.

Answer 6.17

(a), (b) See Figure 6.44.

(c)

(i) $\sigma_{max} - \sigma_{min} = 130 - 30 = 100$ MPa at fluid pressure 20;

(ii) $\sigma_{max} - \sigma_{min} = 120 - 40 = 80$ MPa at fluid pressure 40.

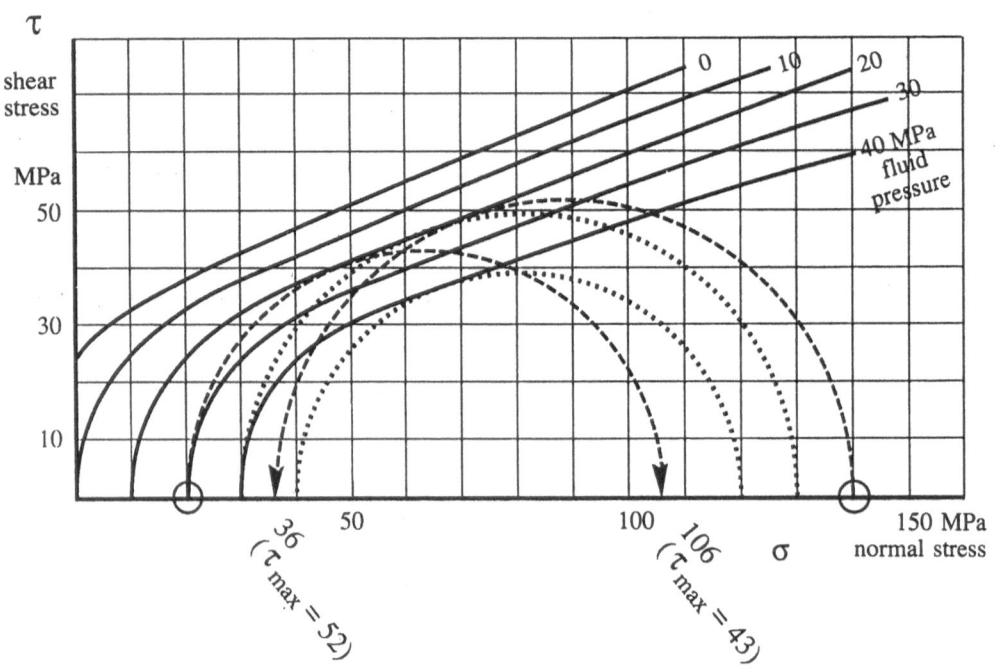

FIGURE **6.44**

Answer 6.18

(a) In Figure 6.45, at normal stress 39 MPa and fluid pressure 10 MPa, shear stress needed to cause slip = 13 1/2 MPa.

(b) $\dfrac{18{,}000 \times 12{,}000 \times 13\frac{1}{2}}{18{,}000 \times 1300} = 125$ MPa average normal stress.

(c)

(i) Consider top of slab first, where σ_{min} (= overburden pressure) = 0. If fluid pressure = 10 MPa, the largest circle that can pass through $\sigma_{min} = 0$ and just touch the curve for fluid pressure = 10 MPa gives $\alpha_{max} = 70$ MPa (see Figure 6.46).

If stress at top = 70 MPa as in Figure 6.47 and we need the average stress at back end = 125

MPa, then, if variation is linear, stress at base needs to be 180 MPa.

At base, $\sigma_{min} = 39$ MPa due to slab's weight, but using the curve for fluid pressure = 10 again, for this value of σ_{min}, the maximum compressive stress the rock can withstand without breaking = 162 MPa; we *cannot* get the average back-end stress up to 125 MPa without cracking the rock somewhere.

(ii) if fluid pressure at top = 0 MPa, σ_{max} at top changes from 70 MPa to 88 MPa and σ_{max} needed at base to give the average value of 125 MPa drops to 162 MPa, which the rock can *just* withstand.

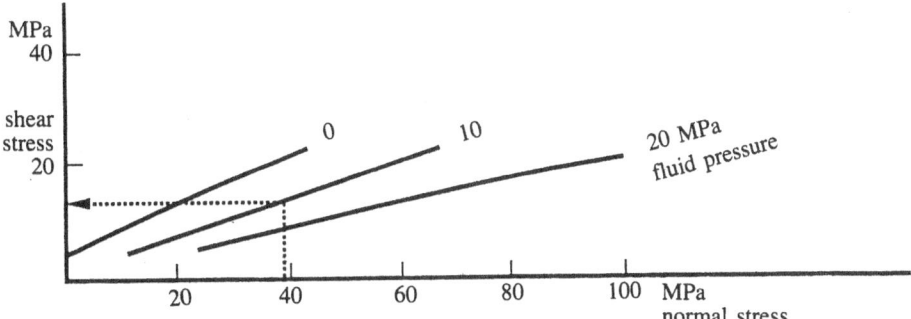

FIGURE 6.45

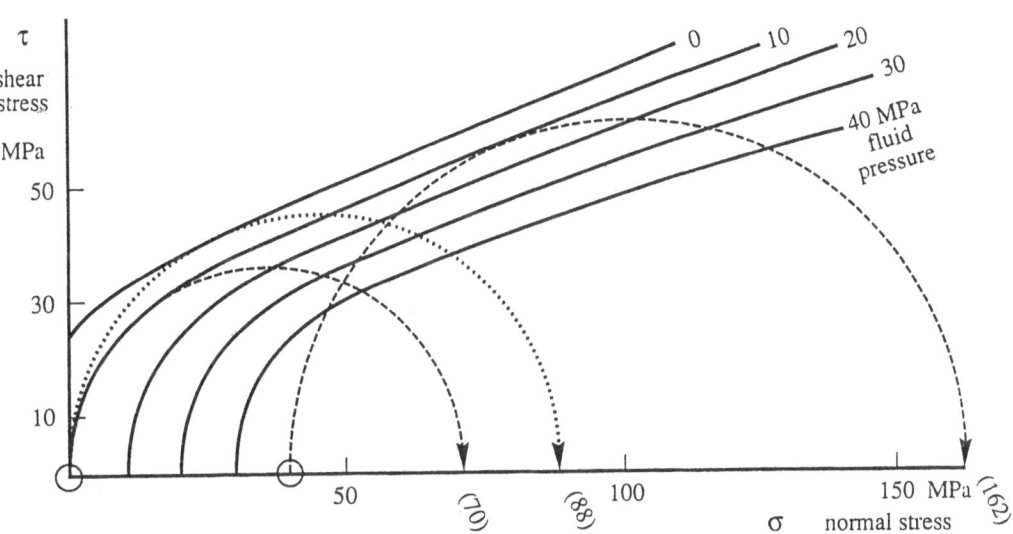

FIGURE 6.46

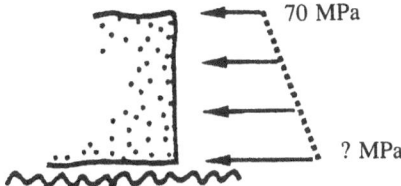

FIGURE 6.47

Answer 6.19

(a) If, as before, we take rock's relative density as 3 giving 3 g/cm³ or 3 t/m³, total weight of slab = $18,000 \times 12,000 \times 1300 \times 3$ t
  - (i) component parallel to slope = $5/13 \times 18,000 \times 12,000 \times 1300 \times 3$ t = 3.24×10^{11} t;
  - (ii) component normal to slope = $12/13 \times 18,000 \times 12,000 \times 1300 \times 3$ t = 7.78×10^{11} t;

normal *stress* = $12/13 \times 39$ MPa = 36 MPa.

  - (iii) Using Figure 6.30, at fluid pressure = 10 MPa and normal stress = 36 MPa, maximum shear stress rock can withstand = 13 MPa = 1300 t/m².

(b) From part (a.iii), maximum *force* parallel to slope

that rock can withstand $18,000 \times 12,000 \times 1300$ t/m² = 2.81×10^{11} t.

This is less than the component of slab's weight by 0.43×10^{11} t.

Dividing 0.43×10^{11} t in the ratio 3/4:1/4 gives 0.11 $\times 10^{11}$ t as the tensile *force* needed at top end;

area = $18,000 \times 1300$ m² so average tensile *stress* = 470 t/m²= 4.7 MPa.

This tensile stress is *less* than the tensile strength of the rock; but the sum (average tensile stress) + (fluid pressure) = 14.7 MPa if fluid pressure = 10 MPa, which is more than the tensile strength of the rock.

Answer 6.20

(a) See Figure 6.48: effective $\sigma_{min} = \sigma_{min} - 20 = 0$ MPa. Starting from 0 MPa and drawing a circle in Figure 6.32 gives effective σ_{max} = 86 MPa. Hence total stress σ_{max} to produce failure = 106 MPa. The shear stress τ_{max} at failure = 43 MPa.

  The single curve in Figure 6.32 used as an average of the curves in Figure 6.29 actually matches the curve for fluid pressure 20 MPa, so these answers are exactly as in Answer 6.17.

(b) Effective σ_{max} = 120 MPa. Drawing a circle

shows that effective σ_{min} = 16 MPa, and hence total σ_{min} = 36 MPa; τ_{max} at failure = 52 MPa.

(c)
  - (i) Effective mean stress = 60 MPa. Then the largest circle with center at 60 MPa gives effective σ_{min} = 10 MPa and effective σ_{max} = 110 MPa; stress difference = 100 MPa exactly as in Answer 6.17.(c.i).
  - (ii) Effective mean stress = 40 MPa and stress difference = 83 MPa, not quite exactly as in Answer 6.17(c.ii), but close.

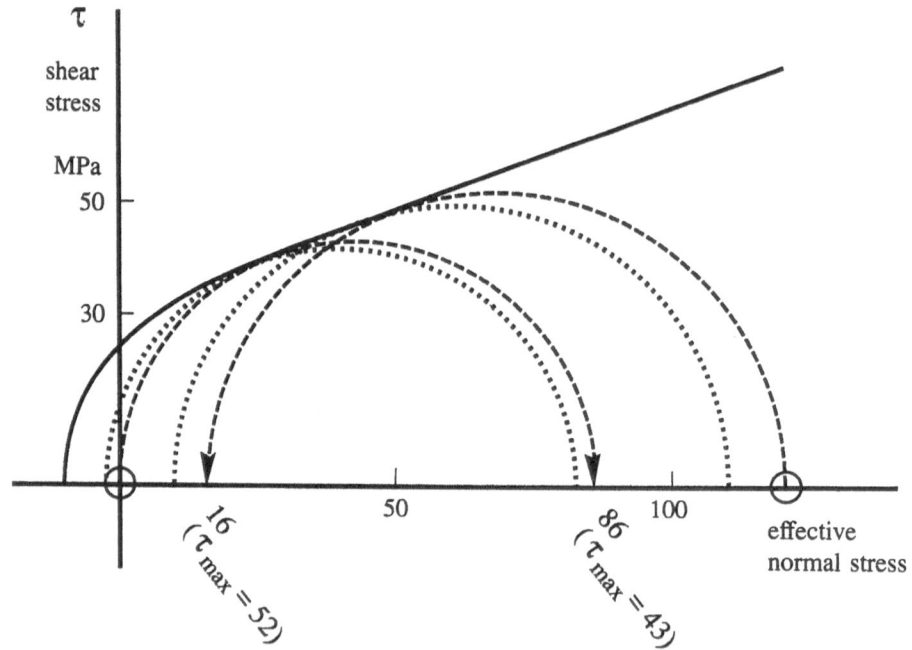

FIGURE 6.48

Answer 6.21

(a) 0.5;   0.75;   0.81,   or 13/16.

(b) The least and greatest local values of the fluid pressure ratio would be 0.72 and 0.88. The effective coefficients are then (0.28 × 0.6), (0.20 × 0.6) and (0.12 × 0.6) or 0.17, 0.12, and 0.07; the largest and the smallest differ from the middle value by roughly 40 percent. In particular, a local fluctuation in fluid pressure by 10 percent upward from the average value takes away nearly half the rock's strength.

Concurrent Fracture and Flow

The idea that fracture and flow can be concurrent is slightly nontraditional. In the simplest of world-views, solids fracture but do not flow, and fluids flow but do not fracture. Observation of outcrops established long ago that this view is too simple; a single layer of rock can certainly show *either* behavior. But the question remains, Can one material do *both*? That is, can fracture and flow be "concurrent"? Is it not more correct to think that when a material flows it doesn't fracture, and when it fractures, it doesn't flow?

LINKS BETWEEN THE PROCESSES

Two ideas help us to see that the behaviors do not exclude each other, and both can be introduced through everyday observations. Firstly there is the flow of salt or sugar out of a bag or jar, or the flow of sand in an egg timer or hourglass. This is a matter of scale; what we call flow on a large scale results from various behaviors on a scale of single grains that are certainly not flow behaviors. The second class of everyday occurrences is shown by taffy or toffee, chewing gum, and various hot plastics and glass: a sample can be pulled out to three or four times its original length but eventually *snaps*. It becomes thinner, it necks down before snapping, but all the same, the snap is a definite, recognizable event, and at the moment before the snap, the material is flowing. In fact it is flowing at the moment it snaps—and sometimes, as one sees when the separated ends droop downward, it continues to flow after snapping, although the driving stress state is of course very different.

With these two everyday behaviors in mind, let us now think of a rock at, say, 15 km depth, where we might find a temperature 400°C, total pressures about 500 MPa, and fluid pressure between 250 and 500 MPa. The maximum and minimum compressions might be 480 and 520 MPa; the rock is not truly red-hot but is perhaps a dull red; what *processes* are going on?

First, the rock will be responding to the stress difference. The fluid content may be in the range of 1/2 percent to 3 percent (or could be less, or more); it will occupy the array of pores and defects mentioned in Chapter 6 (especially under the headings Hydrofracture and Effective Stress). It is likely to be dominated chemically by H_2O or CO_2, and in the former case its relative density will be close to 1: despite the high temperature, it will be more like water than like steam because of the pressure. It will contain chloride as well as carbonate ions and may or may not be acidic—an "acid brine." This is the fluid whose activity as a solvent is shown in Figures 5.24 through 5.28 and 6.25. The compressive stress in the rock is carried mostly by the grains, but partly by the fluid (see Questions 3.18 and 3.23). Because of the stress difference, the grains will be rearranging themselves; their movement will be *hindered* by the strong overall compression but *aided* by the solvent action of the fluid—at any resistive point where local stress concentration gets too high, solvent action will eventually get rid of the obstacle.

The point addressed in the preceding paragraph is that the rock deforms mainly by slip and rearrangement at grain boundaries. The overall process resembles the flow of table salt or sugar, in the sense that most grains retain most of their identity while being rearranged. The term "cataclastic flow" is sometimes used: its strict meaning is the flow of a multi-grain material specifically by fracture processes (cataclasis) acting between one grain and the next, whereas here the focus has been on *solution* processes. A general term would be "grain-boundary-controlled flow." The point is that, while the assembly as a whole *flows*—as smoothly and continuously as a glacier—every grain boundary is acting as a fluid-filled fracture or fluid-aided slip surface. Large-scale flow occurs by means of small-scale fracture so the two effects being concurrent can hardly be questioned.

The second everyday analog (taffy, toffee, gum, etc.) carries the idea that a material is flowing at the moment it fractures, and this follows very readily from what has just been discussed. As just noted, grain-boundary-controlled flow operates at microfractures; if shear stress is large enough and compressive stress small enough, the number of microfractures increases; the effect is self-accentuating in the sense that a site where there are already more microfractures than average will be weaker and will produce new microfractures faster; thus if a through-going fracture develops, it will develop taking advantage of sites where microfractures are already clustered. In short, microfracture and fracture are closely linked; but microfracture and flow are closely linked; hence fracture and flow are closely linked, the fluid-filled microfractures being responsible for both.

Change of Temperature and Pressure   While the vital importance of the swarm of microfractures is in view, it is opportune to note the effect of temperature and pressure. In simple ideal behavior, flow is affected by temperature and is *not* much affected by pressure. But in the world of microfractures or grain-boundary-dominated flow, we find an interesting combination: for some particular stress difference, such as 40 MPa in the conditions sketched above, increasing the overall compression increases grain drag and diminishes mobility—a frictionlike effect; at the same time, increasing the temperature increases the rate of the microprocesses and increases mobility—a fluidlike effect. Engineering provides examples of slip on fractures (insensitive to temperature) and of flow of fluids (insensitive to pressure) but for geology we need a third picture—a grain-boundary-controlled flow that is affected by both; flow goes faster when hotter but goes slower if compression is increased.

In concluding this section, we recall that the discussion started with a picture of the "middle crust" at a depth of about 15 km. Clearly, if we transferred to much deeper conditions, or much shallower, or much drier conditions, the emphasis would change. The program for the next two sections, then, is first to express the preceding ideas in numbers and then to consider how the numerical description will change if we range from shallower to greater depth, and from wetter to drier conditions in the crust.

Summary   Two links between flow and fracture are

1. the effect of scale. What we see as flow on a large scale can occur by rearrangement of a mass of almost rigid grains on a small scale. The overall flow is achieved by grain-boundary processes; on the scale of grains, a changing population of fluid-filled microfractures dominates the action.
2. the effect of local accentuation. A through-going fracture develops out of the population of microfractures. A localized increase in the population of microfractures occurs by grain-boundary processes that add up to localized flow. At least to this extent, a material is flowing at the moment it fractures.

A through-going fracture may move by localized flow that involves microfracturing. In this situation, compression diminishes mobility (a resemblance to friction) while at the same time rise of temperature increases mobility (a resemblance to fluid flow).

If we go from a shallower to a deeper or from a wetter to a drier location in the crust, there may be drastic change in the relative importance of some microprocess, even though the list of processes running remains the same.

A MOHR DIAGRAM FOR FLOW BEHAVIOR

One way of summarizing the influence of stress on rocks is to say that compressive stress holds a rock together, whereas shear stress tears it apart. The saying contains a good deal of truth: the two effects named do compete, and the success with which one or the other predominates determines how the rock behaves. Now, these two quantities are the frame we use for a Mohr diagram, so that that diagram is capable of carrying much more information than is usually put on it. To maximize the convenience, we make a small change as follows.

Stress Points

In a standard Mohr diagram, a point provides information about some single plane. Whatever plane the point belongs to, the point's coordinates show the normal stress and shear stress on that plane, and to show stress magnitudes on all planes through just one material point requires a complete circle of points in the diagram. However, it was emphasized in Chapter 3 that the state of stress at a point can be specified in just two numbers (or three in three dimensions): we can state σ_{max} and σ_{min} or, just as effectively, we can state $(\sigma_{max} + \sigma_{min})/2$ and $(\sigma_{max} - \sigma_{min})/2$, which specify the Mohr circle's centerpoint and its radius. Now these two quantities are the coordinates of a Mohr circle's highest point or crest-point; to show just the crest-point gives the viewer complete information, from which, if necessary, the whole circle could be drawn.

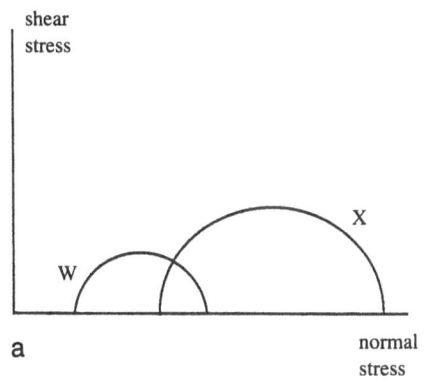
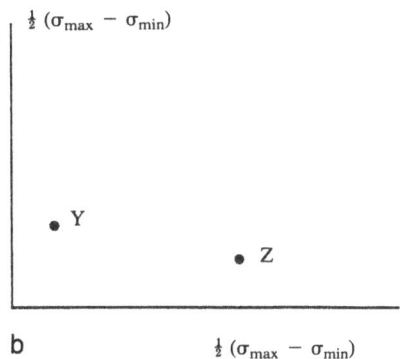

FIGURE 7.1

Question 7.1 Stress circles and stress points:

(a) Two stress states are represented in Figure 7.1(a) by half-circles *W* and *X*. Mark *points* in diagram (b) that represent the same two stress states.

(b) Two stress states are represented in Figure 7.1(b) by points *Y* and *Z*. Sketch circles in diagram (a) that represent the same two stress states.

Failure Envelope Using Stress Points   The function of a failure envelope is to show which stress states fracture a rock. A stress state that fractures a rock is represented by a circle that touches the envelope in a standard Mohr diagram, but is represented by a

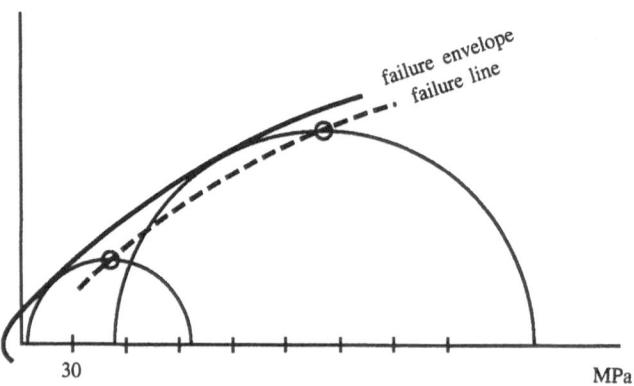

FIGURE 7.2

circle whose crest is on the *failure line* when crest-points are being used. As shown in Figure 7.2, the failure line is just slightly displaced from the envelope.

Question 7.2 Stress circles and stress points, continued:

(a) For the rock illustrated by Figure 7.2, if σ_{min} = 30 MPa, to what value must σ_{max} be raised to fracture the rock? Check that the same answer can be reached by drawing a circle or by finding a suitable crest-point on the failure line.

(b) In Figure 7.2, how would you label the axes if you plan to use Mohr circles and the failure envelope? How would you label the axes if you plan to use crest-points and the failure line?

Question 7.3 Stress points for tensile failure: If a rock has a tensile strength of 8 MPa, it will fail for any of the following pairs of principal stresses: -8, -6 MPa or -8, 0 MPa or -8, $+4$ MPa, or for other stress states like these. Where are the crest points for these three stress states? What is the slope of the failure line for tensile parting conditions?

Flow Behavior Using Stress Points   Up to this point, the use of stress points instead of stress circles has been only a minor change in procedure, but now we come to a significant advantage: any point in Figure 7.2 below the failure line represents a condition a rock might be in, and in that condition the rock is likely to be undergoing continuous deformation at some non-zero rate. Rocks begin to flow before they fracture, and for any point below the failure line, we can add a label stating what kind of flow the rock is undergoing. In practice, of course, we do not label separate individual points; to contour the region below the failure line is more efficient, as in Figure 7.3(a). But the diagrams have most geological usefulness if we change the horizontal axis as in Figure 7.3(b) as follows.

The major change is to conceive the horizontal axis as an axis for depth in the earth. As regards compressive stress magnitudes the change is still minor, because compressive stress and depth increase together. But when the axis shows depth, the right-hand end represents hotter rocks, and so the flow-rate contours take a distinctly different shape. (The contour labels are flow rates per gtu.)

To construct Figure 7.3(b) in definite numerical form one has to select a temperature at

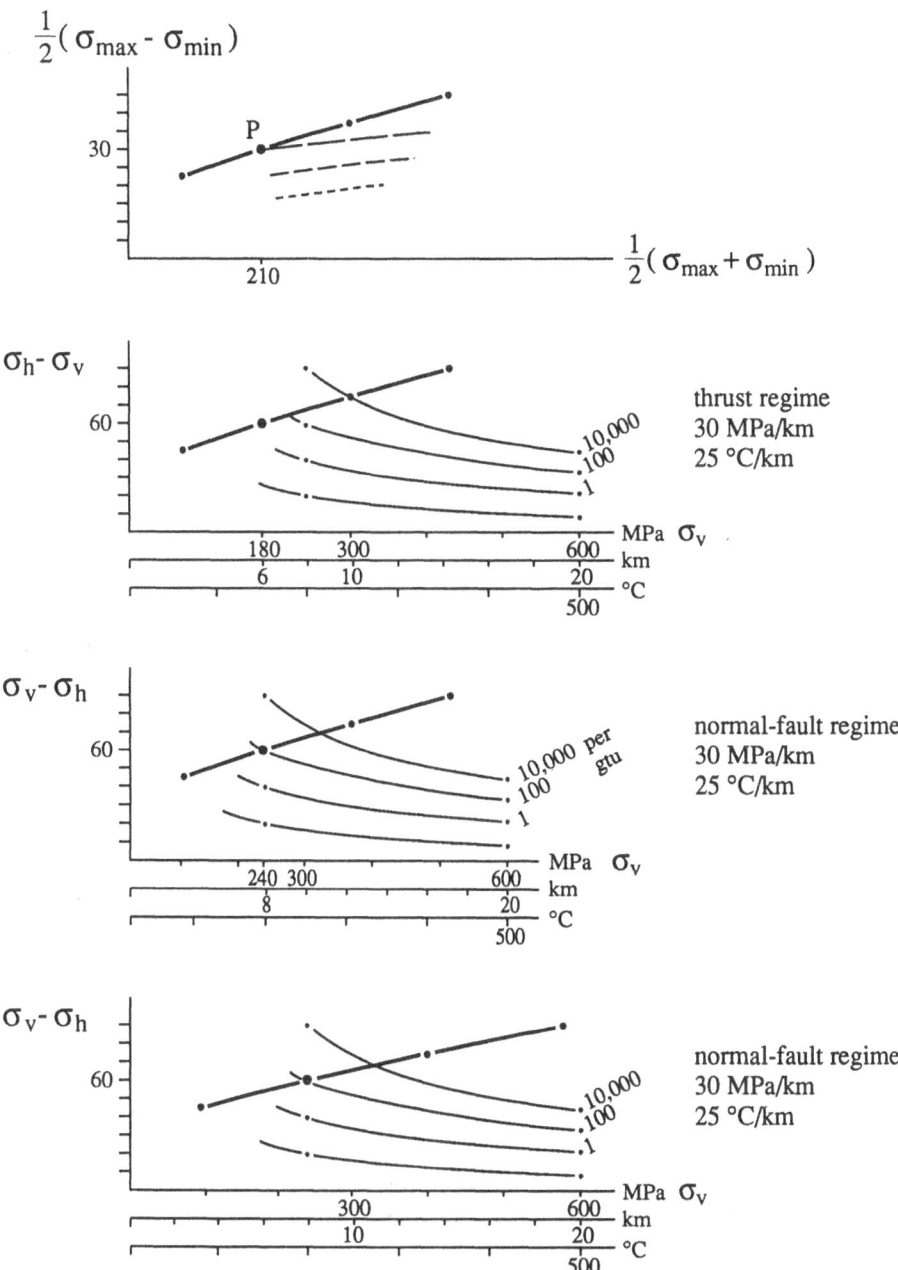

FIGURE 7.3

each depth, for example by assuming that the temperature is 0°C at the surface and increases downward at a uniform rate such as 25°/km. One also has to choose a *regime* —thrust, strike-slip, or normal. The difference this makes is illustrated by Figure 7.3(c). Consider for example point *P* in diagram (a), which shows that the rock illustrated fails at mean stress 210 MPa and stress difference 60 MPa. The maximum and minimum stresses then are 180 and 240 MPa. In a thrust regime, the smaller of these is the vertical stress so that we would look for this stress state at around 6 km depth, whereas in a normal-fault regime, the larger is the vertical stress and we would look for the stress state at greater depth, about 8 km, as in diagram (c). The rock would be hotter in the second state, and the flow rate just below the failure condition would be faster—assuming that the failure condition is not affected by temperature. In reality, the failure condition is *slightly* affected by temperature, but the flow rate is affected much more noticeably. Thus comparing diagrams (b) and (c), the failure line is shown as the same in both but the flow-rate contours are not the same.

Question 7.4 Flow rate at the moment of fracturing:

(a) At 10 km depth, what flow rate does the rock achieve just before failure (i) in a thrust regime? (ii) in a normal-fault regime?

(b) What is the critical magnitude for σ_h in each regime?

The outcome from Question 7.4 is that in the thrust regime, the stress difference is slightly larger and the flow rate is considerably larger (and when a rock is close to fracturing, such sensitivity is not wholly unreasonable). Another way to make the comparison is to use Figure 7.3(d), which is simply Figure 7.3(c) redrawn on a horizontal scale that has been expanded to match diagram (b). Then (b) and (d) show the same flow-rate contours, as is proper: the *rate* of response at some depth and temperature does not depend on whether vertical lines or horizontal lines are shortening. But the failure line passes through the field of flow-rate contours in a different position according to the regime we are in.

Summary

Both flow behavior and fracture behavior can be represented on a diagram whose axes show vertical stress and (1/2 × stress difference).

For a given rock, a separate diagram is needed for each of the regimes *thrust, normal-fault*, and *strike-slip*.

The axis for vertical stress can also be used as an axis for depth in the crust. When this is done, if a geothermal gradient is assumed, temperature also varies along the axis and the flow-rate contours can correctly represent temperature effects.

Each regime and chosen geothermal gradient gives a slightly different diagram, but the overall nature of the relations shown does not change.

As noted in Chapters 5 and 6, deforming a rock affects its deformability: particularly when a rock is close to fracturing, even if the stress state and temperature are kept constant, the rock's effective viscosity tends to drop as deformation continues because of textural changes. The diagrams discussed are *not* well suited to displaying this effect.

A powerful influence on both flow and fracture behavior is the fluid in a rock's pores. The diagrams discussed are very well suited to displaying the effect of a fluid, as discussed in the following section.

EFFECT OF PORE FLUID ON FLOW BEHAVIOR

A rock rich in pore fluid behaves in ways that are vastly different from those in dry rocks: the *amount* of pore fluid present, its *pressure*, and its *chemistry* are all very powerful influences on the ease with which a rock flows. Since mechanics is our topic, and in Chapter 6, the *pressure* of a fluid has already been seen as a key attribute, the program will be first to consider the effect of fluid pressure, leaving amount and composition until later.

Effect of Fluid Pressure on Behavior

The preceding section, and especially Figure 7.3, showed that information about rock flow can be shown in the space below a failure line. Also Figure 6.29 showed several failure envelopes for the same rock at different fluid pressures. If we treated each envelope in Figure 6.29 in the manner of Figure 7.3, overlapping contour lines would get hopelessly confused, but a workable scheme is offered in Figure 7.4.

First Example: Strike-Slip Regime   As in the preceding section, we begin by specifying a regime, a pressure–depth relation and a temperature–depth relation: let us imagine a strike-slip regime with σ_{max} north-south, and use gradients of 30 MPa/km and 25°C/km as in Figure 7.3(b). Then we reflect that fluid pressure is likely to be at least (1/3 × vertical stress) and cannot be more than a few megapascals above total vertical stress, so the range of fluid pressures considered is as in Figure 7.4(a). As the figure shows, ultimately melting sets in, so that if we confine the discussion to rocks unaffected by melting, we are concerned with a triangular field of possibilities. Then the rock's response as σ_{NS} climbs up to exceed σ_{vert} can be shown in serial sections, as in Figures 7.4(b) and (c).

Figure 7.4(b) shows possible processes. It is to be remembered that the diagram purports to show the behavior of some single specimen of rock, of a given composition and texture, but the nonquantitative relationships are likely to be as shown for many common rocks. Major divisions are (1) melting; (2) processes where original grains mostly retain their identity, with action localized at grain boundaries; (3) an intermediate

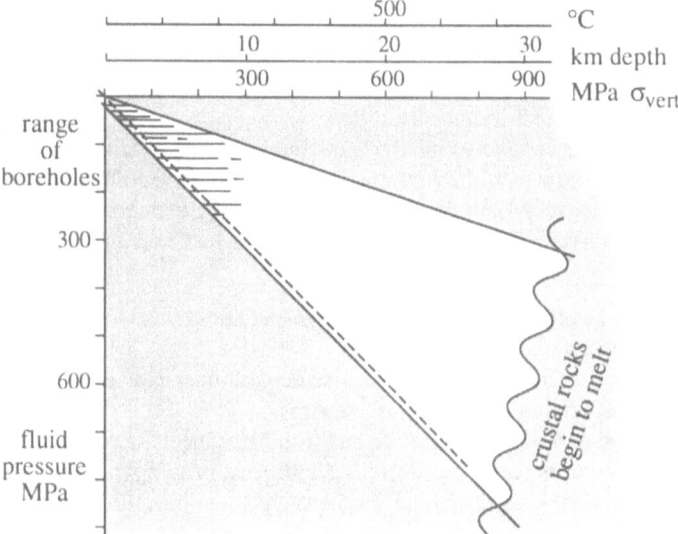

FIGURE 7.4A

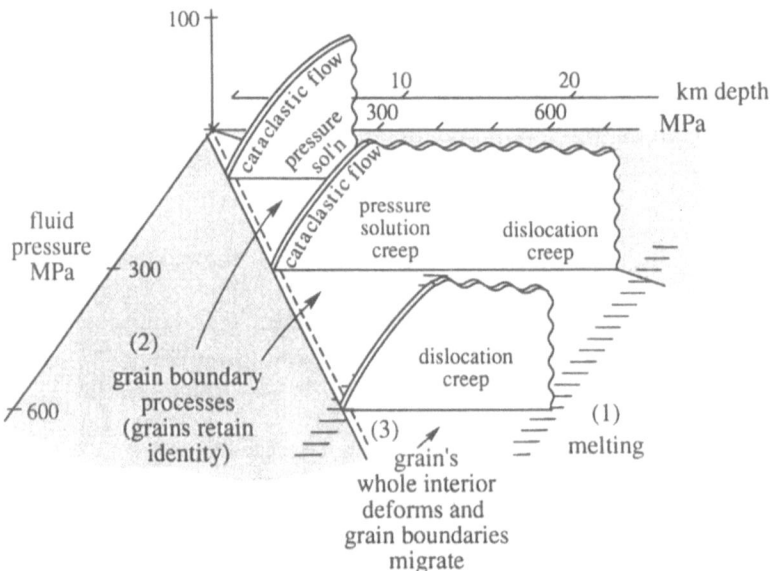

FIGURE 7.4B

region where the grains are sufficiently hot that even the interiors of grains deform, and growth of some grains at the expense of others is rapid. (Because large grains also break up into sub-grains, there may be no overall coarsening.) In field (3), close to melting, the main control on deformability is temperature; the main activity by which a grain's interior deforms goes on at interior surfaces where the crystal structure is imperfect, and the processes are known collectively as **dislocation creep.**

The remaining field, where grain interiors are rather stable and activity is mostly at the boundaries, is covered by two labels **cataclastic flow** and **pressure-solution creep.** As discussed earlier, both processes involve behavior of microfractures. If the effect of the fluid in the microfractures is mostly mechanical, the effect is cataclastic flow, whereas if the fluid is active as a solvent, the effect is pressure-solution creep. The field of cataclastic flow lies up against the failure line because cataclastic flow involves failure; the difference is that if just a single plane fails, we call it a fracture or a shear failure, whereas if the process is more distributed through the rock, with many planes failing simultaneously, we call it cataclastic flow.

In Figure 7.4(c), we first locate the same three fields and, disregarding numbers, note the contour pattern in each. In the cataclastic-flow field the contours have positive slope, in the dislocation-creep field they have negative slope, and in the pressure-solution field they are roughly flat. Recall page 208; the slopes just noted go with the points made there:

Where behavior is frictionlike, a rock's resistance increases with compressive stress and is not much affected by temperature.

Where behavior is fluidlike, a rock's resistance diminishes with rise of temperature and is not much affected by compressive stress.

Where behavior combines frictionlike and fluidlike aspects, a rock's resistance increases with compressive stress but decreases with rise of temperature, with the result that proceeding down a geothermal gradient changes the rock's mobility only slightly.

Another nonnumerical observation in Figure 7.4(c) concerns effective stress. In any of the vertical sections, the distance inward along the horizontal axis from the reference

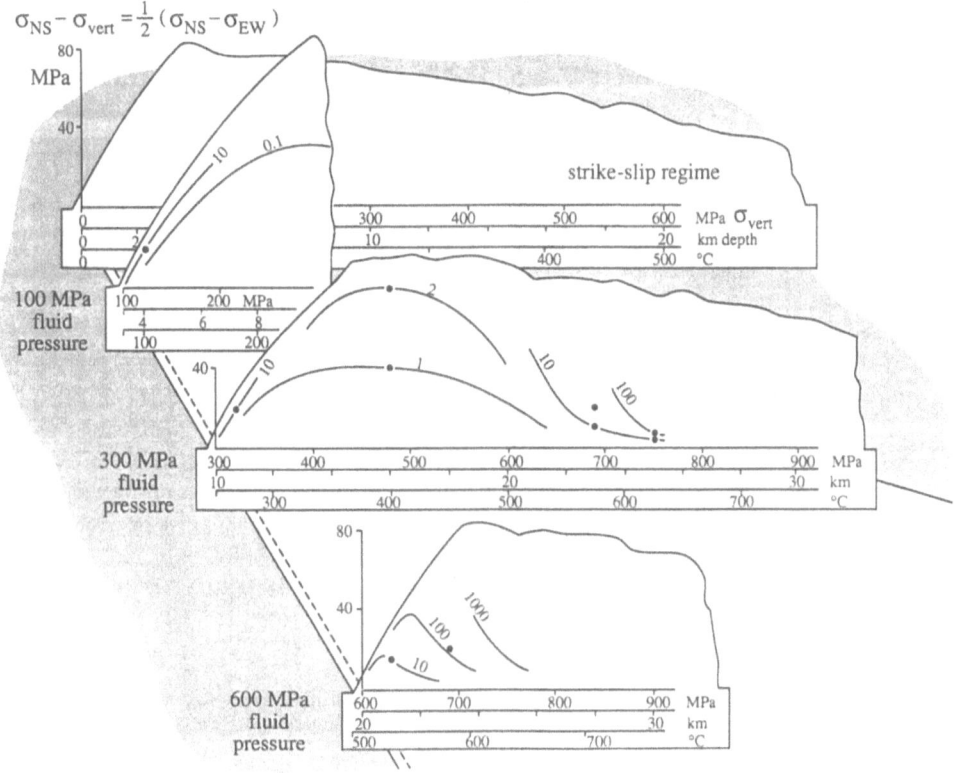

FIGURE 7.4C

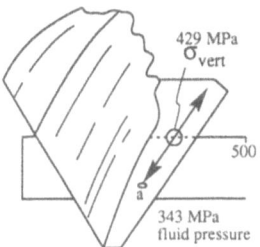

FIGURE 7.4D

point at the bottom left corner shows the effective vertical stress, σ_{vert}-fluid pressure. As long as the effect of the fluid is purely mechanical, the whole rock behavior is well described by just the effective stress. That is to say, in each of the vertical sections, for the first few megapascals in from the left-hand side we see the same contours—σ_{vert} of 320 MPa and P_{fluid} of 300 MPa gives the same behavior as σ_{vert} of 120 MPa and P_{fluid} of 100 MPa. But over all the remaining region of interest, this simple repetition is not found—σ_{vert} of 660 and P_{fluid} of 600 MPa at 22 km depth and 550°C gives very different behavior from σ_{vert} of 360 and P_{fluid} of 300 MPa at 12 km depth and 300°C, even though the effective vertical stress is 60 MPa in both situations. Thus the concept of effective stress can be very misleading if used outside its proper range.

Another warning concerns the idea sometimes put forward that fluid pressure in the earth is a constant fraction of σ_{vert}, such as 0.7. According to this hypothesis, P_{fluid} of

300 MPa would be found at σ_{vert} = 429 MPa and depth 14.3 km. But suppose the fluid pressure varies between 0.6 and 0.8 of σ_{vert}: then at 14.3 km, P_{fluid} would vary between 257 and 343 MPa, and the higher of these values (point a in Figure 7.4(d)) is getting into the field of cataclastic flow. With chains, as is well known, it is the strength of the weakest link that we need to know; the average strength of all links is not a useful number. Similarly with fluid pressure, the average ratio of P_{fluid} to σ_{vert} is not of interest; we need to ask, where does the maximum value of that ratio occur, and what rock mobility goes with it? A single narrow depth-zone where the ratio is locally 0.8 may dominate the behavior of the rock mass, in complete disregard of the main body where the average may be 0.7. In connection with fracturing, a similar point was made at Answer 6.21.

Flow Rates   To complete the discussion of the particular example in hand, we give attention to the flow-rate numbers on the contours shown, which are strains per gtu (1 gtu = 10^{14} sec or about 3.2 million years). As already noted, behavior is affected by fluid *amount* and *composition* and beside that, the numbers are at best only guesses; but there are some expectations that guide us when trying to imagine a possible set of contours. For the numerical work that follows, we assume not only that σ_{vert} lies *between* σ_{max} and σ_{min}, we specifically assume that σ_{vert} = $1/2(\sigma_{max} + \sigma_{min})$.

Question 7.5 Shear stress and flow rate in the crust: It was suggested in Question 5.8 that beneath a fault like the San Andreas fault, at a depth where the rock is too mobile for earthquakes to occur, we might find a shear strain rate of 5 per gtu and a shear stress of 10 MPa, 30 MPa, or that order of magnitude. Using those values as a guide, imagine a location specifically at 21 km depth where σ_{vert} = 630 MPa and P_{fluid} = 600 MPa, and use Figure 7.4(c): what shear stress gives a strain rate of 10 per gtu?

Question 7.6 Shear stress and flow rate, continued: Continuing from Question 7.5, at depth 21 km and P_{fluid} 600 MPa, assume that 10 MPa on the vertical axis gives a strain rate of 5 per gtu. The diagram suggests that for P_{fluid} steady at 600 MPa, if we moved to *shallower* depth, the rock would strain faster and if we moved to *greater* depth, the rock would strain faster again. What reason does the diagram suggest for each trend in behavior?

Question 7.7 Effect of fluid pressure on flow rate: Compare behaviors at 23 km depth for P_{fluid} = 600 MPa and P_{fluid} = 300 MPa.

The comparison in Question 7.7 illustrates the statement made earlier that dislocation creep is more sensitive to temperature than to pressure. The effective vertical stresses considered are 390 MPa and 90 MPa; changing the strain rate from 30 to 150 per gtu requires diminishing the effective stress by nearly 0.8 of its original magnitude. By contrast, a greater change in strain rate is produced by increasing temperature by just 10 percent of its current value; for example, going from 550°C to 600°C increases strain rate by 10× according to the contours shown.

Question 7.8 Effect of shear-stress magnitude on flow rate: At fluid pressure 300 MPa, compare the effect of doubling the shear stress in the pressure-solution field and in the dislocation-creep field.

Before going farther, it should be reemphasized that Figure 7.4(c) is guesswork; the contours display the relationships just discussed merely by artifice. We *start* with a set of flimsy beliefs about the effect of temperature, total pressure, and fluid pressure on rock mobility, and construct contours that conform with those beliefs; then of course it is to be expected that when we read numbers off the contours, the magnitudes will support the beliefs that were originally used.

Summary

Fluid pressure in the earth's crust is likely to be at least $1/3 \times$ total vertical compressive stress and at most a few megapascals greater than total vertical compressive stress (Figure 7.4(a)).

Possible combinations of pressure, temperature, depth, and fluid pressure lead to rock behaviors as shown nonquantitatively in Figure 7.4(b).

To form conjectures about strain rates, one must specify gradients of temperature and pressure with depth, and a regime—thrusting, strike-slip, or normal faulting; then strain rates at different fluid pressures can be guessed at.

Actual magnitudes are not well known, but the general shape of a set of strain-rate contours can be made to conform to the beliefs in Figure 7.4(b).

Experimental work and observation of microtextures, outcrops, and mountain belts each plays a part in helping to limit what the magnitudes might be.

In addition to a fluid's pressure, its composition and amount are also powerful controls on a rock's behavior.

Effect of Fluid Composition and Amount

The attribute of rocks that lies closest to the heart of the present discussion is not their temperature or pressure or any attribute that can be shown in a diagram, it is their *variability*. To understand the mechanics of rock structures, one needs to visit outcrops; an outcrop may teach nothing about mechanics, but it reminds us that one rock is *not* the same as its neighbor. The main defect of a diagram like Figure 7.4(c) is that it provokes silly questions such as: If a rock at 12 km depth behaves as follows . . . , how is the same rock likely to behave at 24 km depth? But we are not likely to be dealing with the same rock at 24 km depth, nor even with a similar rock. Not even "the average rock" will match: the lower crust is different from the upper crust. (But we don't know in what way it is different; some portions of lower crust have been exposed by erosion, but these are an exceptional and nonrepresentative suite.) These reminders are not intended as discouragement—the variety present among rocks makes their mechanical behavior a rich and interesting topic; the reminders serve first as a caution on the use of Figure 7.4(c), and second as a peg on which to hang a further thought: the pore fluid is chemically as complex and variable as the rock host.

Fortunately this fact does not increase the burden on one's imagination. Give Figure 7.4, it is necessary, as just noted, to treat the numbers suggested with caution: if the rock type were different, the diagram would be different though of basically the same type. And if the rock were the same but the *fluid composition* were different, again the contour shapes and numerical values would change—water in a deep mine is often strongly acid, whereas in East Africa some volcanoes exude sodium carbonate and the associated pore fluid is strongly alkaline; the solutes and the quantities a pore fluid can carry are clearly variable from site to site. But no new patterns of thought are needed to accommodate this fact.

When we turn from a fluid's composition to its amount, something more novel needs to be introduced, or reintroduced. As noted in Chapters 5 and 6, deforming a rock affects its deformability, and change of fluid amount is one of the critical factors. As elsewhere, a simple numerical discussion can be given. The purpose is not so much to describe the natural world as to make one's imagined picture a little more detailed and less cloudy. A reason for fluid amount being given a separate discussion is that, whereas composition does not change rapidly through time, the amount of fluid at a site can change radically even in a geologically short moment. The following paragraphs illustrate this.

A Changing Population of Microfractures   Two ideas that have been used already are (1) that a rock in the earth's crust is full of fluid-filled defects, particularly on a scale about the same as the grain size, and (2) that whereas compressive stress tends to heal the defects, shear stress tears them apart. As usual, a realistic description is too complicated to express in symbols; again we seek an idealization simple enough to be handled, yet retaining the essence of the behavior we wish to understand. Here the idealization is as follows: we imagine that all the defects are identical—a "standard representative defect"—and that the number present per cubic millimeter is the variable that, by increasing or decreasing, makes the rock weaker or stronger.

Let the number of weaknesses per cubic millimeter be w, and let the stress state be described by a mean stress of σ and a stress difference $1/2(\sigma_{max} - \sigma_{min})$ of τ. Then idea (2) can be expressed as

$$\frac{dw}{dt} = A\tau - B\sigma. \tag{7.1}$$

The rate at which the rock gets weaker is the rate at which weaknesses accumulate, dw/dt; shear stress τ makes the number of weaknesses increase, and compressive stress σ makes it decrease; at any moment the net change is a result of competition between these two effects. But it is oversimple to suppose that A and B are *constants*: it is more realistic to make A and B vary with w itself, as follows.

The term $B\sigma$ represents the number of weaknesses in a cubic millimeter that are healed in 1 gtu by compression; surely this number will be larger if many weaknesses are present and smaller if there are only a few left to heal; so we make B proportional to w. As regards the term $A\tau$, we use everyday experience: when tearing wood, pastry, wet paper, or cloth, it is difficult to get a tear started in flawless material, but once a sample has many flaws and is only hanging together by a few shreds, the remaining resistance is easily overcome. That is to say, for a constant shear stress, the rate of weakening goes up as weaknesses accumulate. We express this algebraically by making A proportional to $1/(k - w)$; the constant k is large so that when w is small this is a small factor; but as w increases the factor increases, and when w gets close to k the increase is dramatic. The constant k can be thought of as the number of weaknesses that needs to be present for the rock finally to give way.

The expression that results from these thoughts is

$$\frac{dw}{dt} = \frac{j}{k - w} \cdot \tau - hw \cdot \sigma, \tag{7.2}$$

where h, j, and k are all constants, and we imagine subjecting a rock to fixed stresses τ and σ. Does the rock slowly get stronger through healing, or does shear stress tear it apart?

Question 7.9 A changing population of microfractures:

(a) Let σ = 400 MPa, τ = 8 MPa, k = 20 per mm³, h = 3 per MPa-gtu, and j = 6000 per MPa-gtu-mm⁶.
   (i) What is dw/dt if the current value of w is 2 per mm³?
   (ii) What is dw/dt if the current value of w is 8 per mm³?

(b) Repeat parts (i) and (ii) of (a) for a larger shear stress, τ = 24 MPa. Calculate also a rate dw/dt when w = 16 per mm³.

(c) Consider the separate magnitudes of the two terms whose difference forms dw/dt. What is the overall history of w first at τ = 18 MPa and then at τ = 24 MPa? (*Note:* 1000 years $\cong$ 1/3200 gtu; it is useful to write dw/dt as a change in w per 1000 years.)

The behavior in the exercise is illustrated in Figure 7.5. The essential ideas are that

1. in some conditions a rock's texture is stable; change brings it closer and closer to a steady deforming state;
2. in some conditions a rock's texture is unstable; change brings on more change; the process is self-accentuating and, once started, is hard to stop.

Beside being self-accentuating, the runaway tendency is also self-localizing: a rock mass always has at least a little variation in texture and if, for example, five zones or layers differ from one another slightly, the one that gets into runaway behavior earliest will capture the action. It quickly becomes *much* weaker than the others, and a process that begins as slow deformation throughout the mass converts to rapid deformation in a narrow zone. What begins as flow can later be described as a shear failure or even a fracture.

Fluid Amount and Microfracturing   The essence of the present chapter is that flow processes and fracture processes go on together; Chapter 5 and Chapter 6 are only preparations for considering how rocks behave. And the essence of the present section is that a changing content of fluid can rapidly shift the balance or alter the mix. There are two links between fluid amount and the runaway behavior just discussed, as follows.

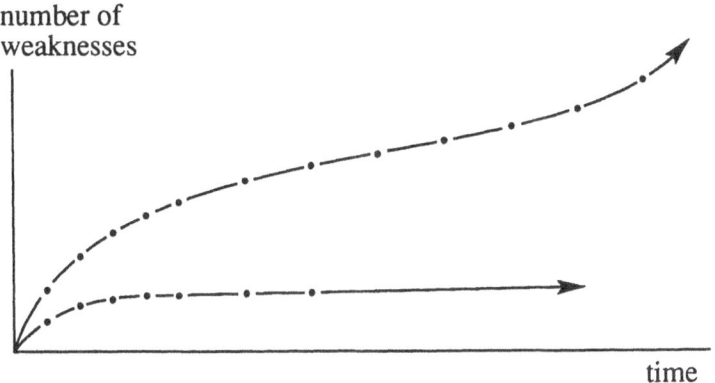

FIGURE 7.5

The first link is that each weakness is likely to contain fluid (a water-rich or CO_2-rich brine) so that as the number of weaknesses or microfractures increases, the rock fills up with fluids. One sees the effect when walking on wet sand: as each foot descends, it deforms the sand immediately beneath it; the deformation increases the number or size of the fluid-filled spaces between the grains, as in Figure 2.6, and fluid is sucked *out of* the less deformed material around each footprint. Stepping on a wet sponge makes the surroundings get wetter but stepping on wet sand makes the surroundings get drier. The increase in fluid is here the *consequence* of deformation; a runaway shear failure picks up fluid.

The second link is that increase of fluid can *cause* deformation; i.e., it can destabilize a texture that was stable. It is easy to see this through the factors h and j in Equation (7.2). To increase j in comparison with h is just as destructive as to increase τ in comparison with σ; j measures the ease with which shear stress tears a rock open whereas h measures the ease with which compression heals it; it is easy to imagine that putting more fluid into a rock increases j at the expense of h.

We need to add an idea from Question and Answer 6.19: the fracture or shear zone a fluid initiates often allows the fluid to escape. This completes the train of thought that prompted the discussion of microfractures in the preceding section: a rock's fluid content can change its behavior more *rapidly* than any other factor. The idea of pulses reappears: as long as intergranular fluids are scarce, a rock's flow can be slow and steady, but if the amount of intergranular fluid increases, deformation does not simply become faster—it definitely tends to become less steady and more jerky. A rock fails when, by change of texture, its strength drops down to equal the stress currently acting: if we read from Figure 7.4(c) that at 16 km depth, 300 MPa fluid pressure and 40 MPa of stress difference, a rock's creep rate is 1 per gtu, and we imagine this to be a placid process because the failure line is far away, we could be wrong. A change in fluid amount could quite rapidly bring the failure line down to where we are; the rock would give a jerk and let the fluid escape, and placid flow could resume until the next jerk; but the cumulative effect over time would be a combination of steady and unsteady, fracture and flow, sticking and slipping. If we try to understand rock structures, we have to handle a lively and versatile material.

CONCLUDING EXAMPLE: THE PYRENEES

Part of the richness of geology is the range of scales. Mountains are built by the movement of grains; one discovers with a microscope the means by which the Himalaya rose so high. In most of the preceding chapters, single ideas from mechanics are illustrated by both small-scale and large-scale examples. But the present chapter is different: here we no longer jump from a small-scale example and take up a large-scale example instead; here we keep both scales in mind at once. We note for example that large-scale "continuous" flow occurs by action at small-scale discontinuities. A consequence is that the mechanics gets to be *variable;* so can we still proceed? One might summarize the preceding section by saying that the behavior depends on the fluid content and the fluid content depends on the behavior; if we have no firm information about either, the problem begins to seem intractable. The purpose of this concluding section is to consider, Are things really so bad? In earlier chapters the Pyrenees have been discussed in several over-simple ways. If we try to bring more sophisticated ideas from the previous section to bear, can we still make progress?

The conclusion is that there is much work still to be done. On the other hand, the component parts *are* tractable. The web of interconnections may become intricate, but computers are well adapted to coping with that. A person who has done the exercises through Question 7.9 has all the needed ideas in mind. It remains only to *keep* them in mind simultaneously, with the proper interconnections. The following is just a preliminary sketch, intended to encourage the reader along such paths.

Assembling the Parts

Exercises that have dealt with the Pyrenees are Questions 2.11, 3.32, 5.9, and 5.10. These introduced quantities such as

width	initially 200 km, but diminishing
crustal thickness	initially 40 km, but increasing
width diminishes at	30 km/gtu
theoretical steady rate	9 mm/yr.

In the strip that is confined between the more rigid blocks of France and Spain:

average strain rate for shortening horizontally and elongating vertically	0.17 per gtu
speculated driving stress difference $\sigma_{horiz} - \sigma_{vert}$	80 MPa
time-averaged apparent viscosity	120 gvu
viscosity if deformation is concentrated in bursts that occupy one-tenth of total elapsed time	12 gvu

In the uplifted sideways-spreading cap:

speculated height above the flanking continents	more than 4 km
time-averaged apparent viscosity	120 gvu (as above)
viscosity for action in bursts	12 gvu (as above)
confining horizontal compressive stress	84 MPA
strain rate for sideways spreading, shortening vertically and elongating horizontally	0.075 per gtu
width increases at	11 km/gtu
total displacement rate of spreading cap relative to closing jaws (both sides)	41 km/gtu

The numbers given suggest that for the triangular wedge in Figure 7.6 (which is half of Figure 5.5 repeated), the displacement rate at the sliding surface is 20 or 21 km/gtu, and the horizontal compressive stress at the heel of the wedge is 84 MPa. We compare these numbers with the viscous hypothesis illustrated by Figure 5.12 and with the sliding-

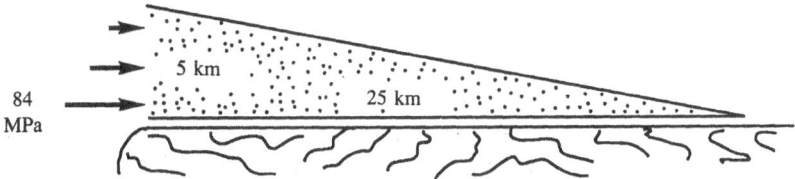

FIGURE 7.6

wedge hypothesis of Chapter 6, and then consider how the whole package is affected by the preceding section. The numbers listed are merely rough guesses or *possible* values and may need to be changed. But the connections among them are based in mechanics: it is the connections among the numbers rather than the numbers themselves that form the underpinnings.

Question 7.10 Viscous sole:

(a) If horizontal compressive stress acts on the back end of a wedge as shown in Figure 7.6, and the shear stress at its base is uniform from heel to toe, what is the shear-stress magnitude?

(b) If the sole is a 100-m-thick rock layer of low viscosity, what must its viscosity be to allow slip of 20 km/gtu when driven by the stress in part (a)?

Question 7.11 Frictional sole:

(a) If the shear stress at the wedge's base diminishes linearly from heel to toe, what is its magnitude at the heel?

(b) What effective coefficient of friction will allow the wedge to slide?

(c) If the rock's regular coefficient of friction is 0.6, what ratio of fluid pressure to overburden pressure would allow the wedge to slide?

Neither of the preceding answers is satisfactory. In Figure 7.4(c) at 5 km depth and 100 MPa fluid pressure, a contour for strain rate 200 per gtu could theoretically be squeezed in; but it would be so close to the failure line that to call the process viscous rather than cataclastic would be misleading. Going beyond Figure 7.4(c), we should consider other rock types: could a rock be so rich in wet clay as to behave viscously rather than as in the figure? We don't find clay seams 100 m thick, and where clay, rock flour, or gouge does occur at a thrust plane, it is likely to have been *produced* by cataclasis. All these thoughts steer us toward the second idea, a frictional sole; here again we find a difficulty, but perhaps it is curable, as follows.

The difficulty with Answer 7.11 as stated is that a fluid pressure of 8/10 × total vertical compression amounts to more than 110 MPa. If the horizontal compression were only 84 MPa, the fluid would split the overburden and escape through vertical fissures. To avoid this effect, the horizontal compression must be a larger fraction of the vertical compression. For example, if at the heel vertical compression is 140 MPa and horizontal compression is 112 MPa, Answer 7.11(a) would be 22.4 MPa and an effective coefficient of 22.4/140 or 0.16 would serve. This is 0.27 × the regular coefficient, so that a fluid pressure 0.73 × vertical stress would let the slab slide and would *not* create vertical fissures.

We conclude that the shallow-level spreading in Answer 5.10 is an overestimate: the shallow-level stress state must be closer to lithostatic than was there supposed. Of the slip that occurs under the wedges, it seems that most is due to inward movement of France and Spain and little, if any, due to outward spreading of the up-bulge.

Source of the Fluid

It appears from Questions 7.10 and 7.11 that magnitudes can be found for the quantities involved that make the story hang together in a coherent way. But the fluid at Question 7.11 is a necessary component and the likelihood of a real fluid being available deserves attention.

It has been emphasized that the mechanical effects of a fluid are often transient. It is possible for a fluid to assist in a slow, steady deformation, but probably most of the deformations that are fluid-assisted are jerky motions: fluid pressure rises, motion ensues and the motion allows fluid to escape. Recurrent motion then requires a source that builds the fluid pressure back up, to initiate the next jerk. The preceding section was mostly arithmetical, shifting the numbers around until they match up; but the present section raises a more geological question, where do the fluids come from?

Along ocean trenches, wet sediments build up accretionary prisms and much water is released during compaction of the sediments themselves. Subducted sediments are another source: Andean volcanoes, for example, are supplied with water that travels hundreds of kilometers as a part of a subducted slab before being released and percolating upward. But in the Pyrenees, the most obvious source is the down-bulge: the down-bulge supplies water to the up-bulge.

In Figure 3.22(c), if the crust is 40 km thick and if, in thickening to 50 km, there is 2 km of upward motion and 8 km downward, we can think of the top 8 km thickening upward, the lower 32 km thickening downward, and a horizontal line at depth 8 km remaining horizontal. A rock originally at, say, 12 km depth and 300°C then gets hotter: its "blanket" is thickening at 12 km $\times$ 0.17 per gtu (the strain rate comes from Answer 2.11) or 2 km per gtu. Let us accept the rough idea that the temperature gradient of 25°C per km is not affected by the deformation: then after 1 gtu, a rock that started at 12 km and 300°C will be at 14 km depth below the rising surface and hence perhaps at 350°C.

Let us suppose further, for the sake of an order of magnitude, that the descending rock contains 30 percent chlorite. On total dehydration, this quantity of chlorite can liberate 8 percent of the rock's volume as water and the dehydration might be spread over the interval 300° to 500°C. If heating from 300° to 350°C liberated one-fourth of the total available, this amounts to 2 percent of the rock's volume; but the rock below, which gets heated from 350° to 400°C, will also liberate the next 2 percent of its volume, etc. The net flow of water through the 300°C isotherm could be 8 percent of 2 km per gtu or 160 m per gtu: as 2 km of rock go down, this much water rises through it. If one imagines a layer of water inside the earth that is 160 m deep, that seems like an unlikely occurrence but the following thought gives a more reasonable picture.

Suppose the deformation is jerky, that each jerk gives a large earthquake and that large earthquakes occur every 330 years. That gives 10,000 jerks per gtu and at each jerk, 1.6 cm of water has to be lost, from across maybe one half the total horizontal area of the mountain belt. (It is only the central part that descends at the rate described.) If a 50 km width descends, the descending area just equals the area of the soles of two thrust sheets 25 km wide, so that we have to imagine 1.6 cm of water being pumped into the rock layer or fault zone that forms the sole and then escaping again. That quantity of water could drastically affect the mechanical properties of a sole-layer 10 cm thick, or somewhat affect a layer 1 m thick; if the active layer were 100 m thick, it is not so easy to imagine 1.6 cm of water making a significant difference.

The preceding numbers are clearly of only the roughest kind, but they illustrate two points: (1) the volume of water available in the deforming mountain belt is large enough to be mechanically significant, but (2) the duration of mountain building is so long that it is difficult to get much of the water accumulated at any one time. The numbers also illuminate our degrees of ignorance: we have a good picture of the probable vertical stress and at 12 km depth we have some basis for estimating an average strain rate; but the relations among fluid pressure, fluid content, and rock permeability (which fix the duration and intensity of a fluid spasm) seem at present largely outside our grasp.

A Unified Mechanics

In the preceding section, attention was given to fluid pulses, and to bursts or spasms of deformation. These thoughts bring us back to the point emphasized early in this chapter: that flow and microfracturing are hardly separable, and that *localization* of flow grades into fracturing on a larger scale. Geological examples of "simple flow" on one hand and "simple fracture" on the other can be found, but in much of the crust such a separation is not clearcut. Yet the mechanics in Questions 5.9 and 5.10 and in 7.10 and 7.11 has continued to be of the two simple kinds. What needs to be changed?

It seems that Chapters 5 and 6 need combining; we need some train of thought—if possible one that can be expressed in numbers and equations, or at least in diagrams— that combines the essences of Chapters 5 and 6. At this point we leave what is familiar and well tested, and embark on untested speculation, but let us do it nonetheless.

Question 7.12 Combining concepts that have up to now been separate:

(a) In what way can Chapter 5 be modified to incorporate ideas from Chapter 6?
(b) In what way can Chapter 6 be modified to incorporate ideas from Chapter 5?

Two suggestion can be made in response to Question 7.12. A strong thread in Chapter 6 is that an increasing normal stress decreases a rock's response to shear stress. Chapter 5 can incorporate this idea if we make a rock's viscosity increase as normal stress increases; but to allow for the effect of a pore fluid, viscosity should be linked first to the *effective* normal stress, $\sigma_{total} - P_{fluid}$. Similarly a strong thread in Chapter 5 is that a small shear stress or stress difference $\sigma_{max} - \sigma_{min}$ gives a slow response, and a larger stress difference gives a faster response, but still a response related to the stress-difference magnitude. Chapter 6 can incorporate this idea if we make a rock respond to shear stress, not by total failure but by something more restrained; we need to imagine that even after "failure", a relation exists between stress magnitude and displacement rate. The simple behavior in Chapter 6 is portrayed in Figure 7.7(a); only small changes are needed to introduce diagram (b); but diagram (b) is of a *type* that grades into more variable behaviors such as the behavior shown in diagram (c).

A convergence can now be noted: we reach roughly the same endpoint whichever of the two chapters we start from. Figure 7.7(b) shows a material that has a small range of critical stress-difference magnitudes; below this range its response to a stress difference

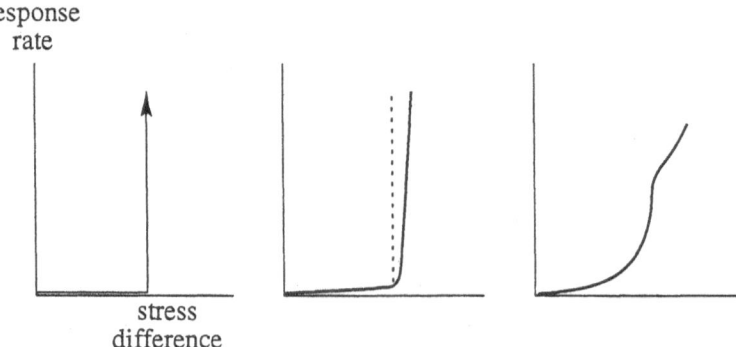

response
rate

stress
difference

FIGURE 7.7

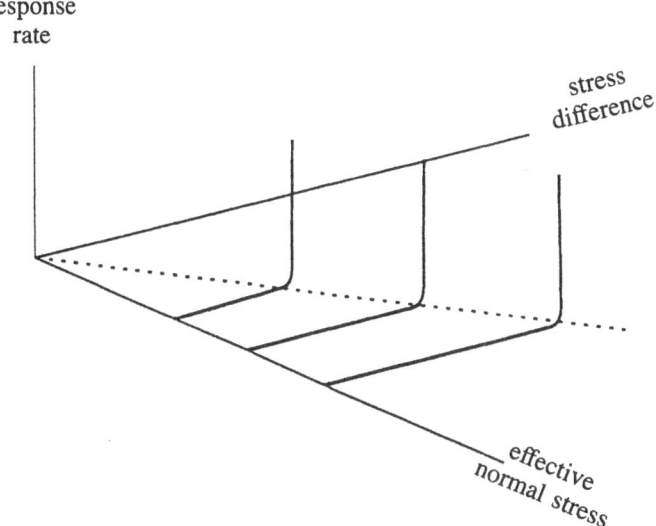

FIGURE 7.8

is very slow whereas above the critical range, its response to a stress difference is much faster; and we bring from Chapter 6 the idea that at a higher effective normal stress, the critical change comes in at a higher stress difference, as in Figure 7.8.

Along the other route, emphasizing viscosity, Figure 7.9 shows three simple viscous behaviors. In each, the response rate is directly proportional to the stress difference, as if the material's viscosity were fixed, but the viscosity is made larger at higher normal-stress magnitudes. It seems that most of Chapters 5 and 6 can be combined if we envisage behavior of the type shown in Figure 7.10.

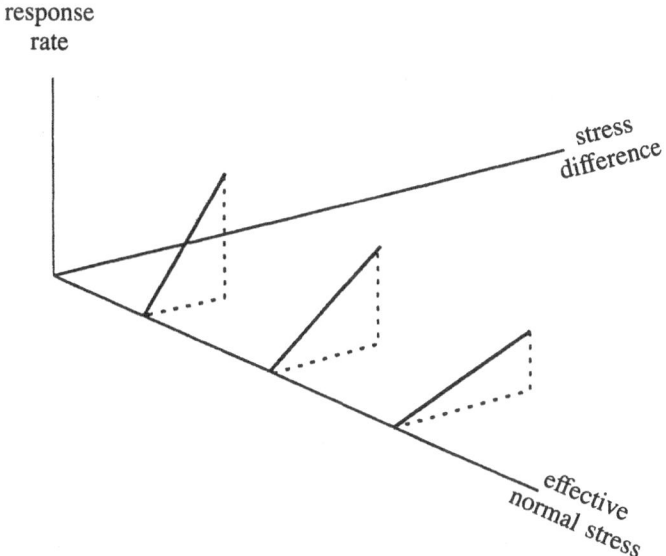

FIGURE 7.9

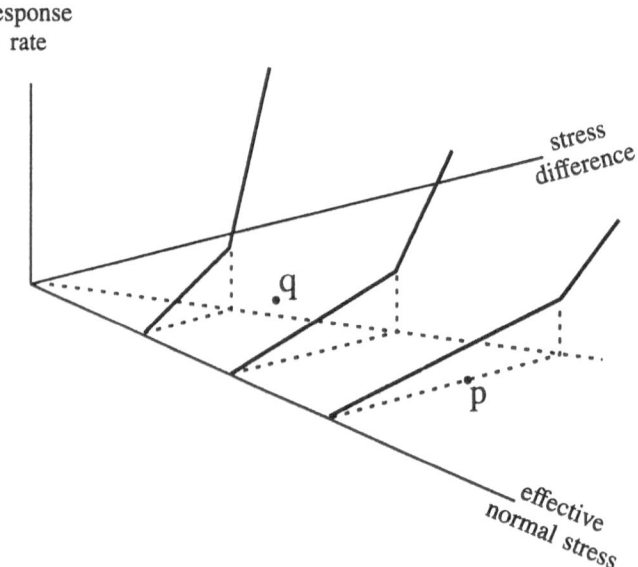

response
rate

stress
difference

q

p

effective
normal stress

FIGURE 7.10

The difference from Figure 7.9 is that we have given up simple proportionality between stress difference and response rate; the difference from Figure 7.8 is that the well-defined strip of high curvature is not so strongly marked. But in concept, a diagram of the type shown in Figure 7.10 seems capable of combining the essences of Chapters 5 and 6, and can of course be modified to put the emphasis on one characteristic or another.

Fluid Pulses and Dilation   The essence of Figure 7.10 is that a rock's response is a displacement rate or deformation rate that increases nonlinearly with stress difference in a manner that is affected by the effective compressive stress (and since the *weakest* plane is the most critical, we watch particularly the *minimum* effective compressive stress). How well can the diagram portray the effect of a pulse of fluid entering the rock?

To some extent the diagram has the needed versatility, because the front axis shows *effective* stress. Suppose conditions are represented by point *p* before a pulse of fluid arrives: one effect of the pulse is to lower the effective compression (without immediately affecting the driving stress difference), taking us to a point such as *q*. Clearly the response rate jumps up, and then after the pulse passes, behavior reverts to point *p*.

The use we make of this type of diagram depends on the scale of the problem at hand. Over a complete mountain belt and more than 1 gtu, we need a diagram that *averages* behavior. On the other hand, as just discussed, on the scale of a single shear zone and a few hundred years, a diagram of the same basic type can give a more detailed picture. But a caution is repeated from earlier pages: deforming a rock affects its deformability, and specifically, *dilating* a rock can be highly destabilizing. The amount of fluid that enters a localized weakness is as important a factor as its pressure, and the fluid amount/fluid pressure/permeability tangle is not an easy one to sort out. Overall, Figure 7.10 is a useful peg on which to hang thoughts, but until fluid amount can be woven in, it gives only the beginnings of what we need.

response
rate

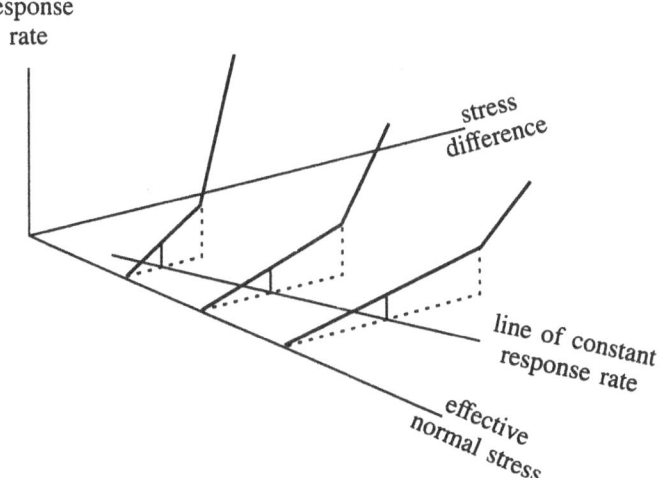

stress
difference

line of constant
response rate

effective
normal stress

FIGURE 7.11

Question 7.13 Comparison of two diagrams: How does Figure 7.10 link up with Figure 7.4(c)? Are the two figures totally independent, or are they two different representations of the same ideas, or is there some kind of partial overlap?

There is almost total overlap. As just noted in the text preceding the question, Figure 7.10 enables us to trace the effect of a transient pulse of increased fluid pressure: if the driving stress difference stayed constant, the response rate would climb, or if the response rate stayed constant, the stress difference needed would drop. Figure 7.11 shows the latter effect: for some specified response rate, a line on the floor of the diagram shows a series of states that would give that rate. Now this is what each contour in Figure 7.4(c) shows, except for one important difference: in the present discussion we are imagining that only fluid pressure changes, while temperature, total pressure, and depth remain unchanged. In other words, the contour in Figure 7.11 would appear on a wall or fence as shown in Figure 7.12(a). And indeed, Figure 7.4(c) permits contours in such planes to be constructed, with results that are of the type proposed in Figure 7.11; see Figure 7.12(b).

Review and Summary

Figure 7.12(b) grows out of Figure 7.3(a), while Figure 7.4(c) grows out of Figure 7.3(b) or (d). The two approaches are both useful, and do not conflict: the contours in Figures 7.4(c) and 7.12(b) together form a set of nested surfaces, as Figure 7.13 shows in an incomplete way. Having assembled several quantities into this one diagram, it is time to think back to the actual rocks, and recall what it is that the diagrams are intended to represent.

A sample of rock has a temperature, a porosity, and a fluid pressure, and is subject to three principal stresses. We approximate the effects of the three stresses by two quantities, σ_{vert} and $\sigma_{max} - \sigma_{min}$, the relation between these two differing according as we

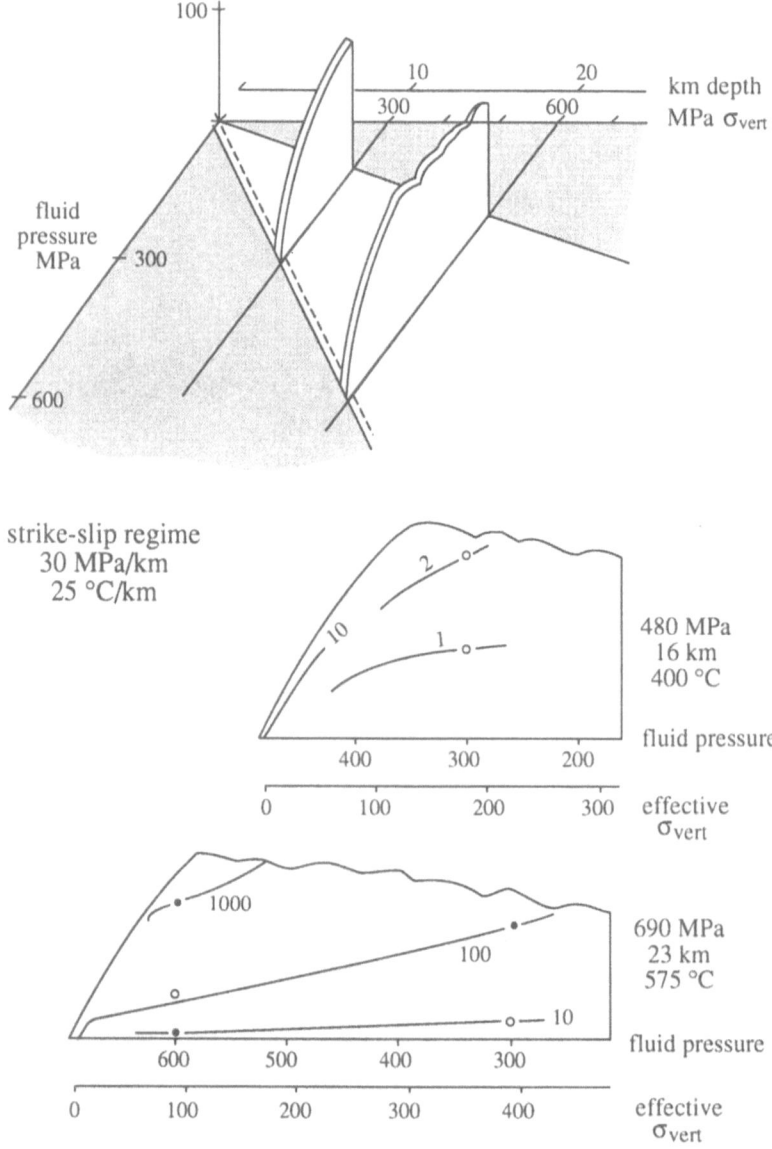

FIGURE 7.12

envisage a strike-slip, normal, or reverse-fault regime. We try to imagine how the rock will respond, maintaining three beliefs:

1. At low temperatures and low effective stress, behavior is dominated by fractures or microfractures and is greatly affected by normal-stress magnitude (see left-dipping surfaces in Figure 7.13).

2. At high temperatures, behavior is dominated by dislocation creep and grain-boundary migration; the response rate is sensitive to temperature and not much affected by normal stress (see right-dipping surfaces in Figure 7.13).

3. Many structures seen in outcrop probably formed in conditions not well described by 1 or 2 above; they formed in an intermediate state where grain-boundary processes (slip

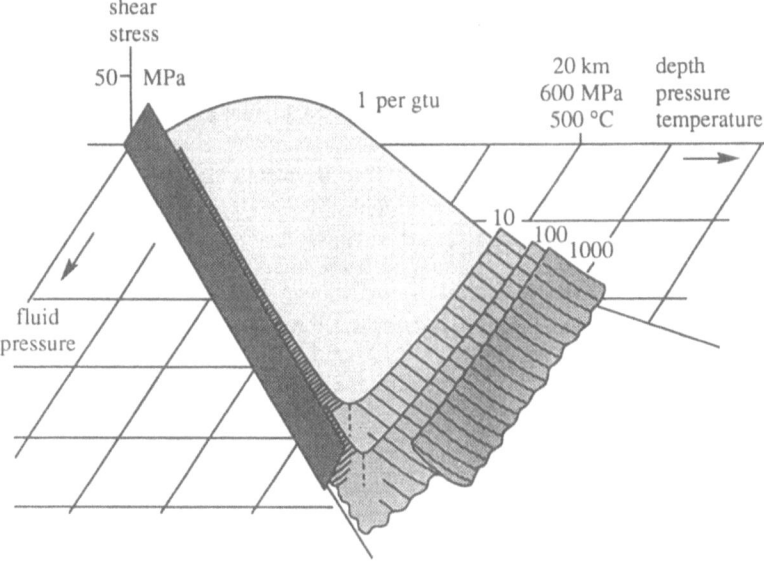

FIGURE 7.13

and solution transfer) played important roles. Behavior can be thought of as controlled by two parameters, depth and effective stress. With increase of depth, rise of temperature increases a rock's mobility but rise of compressive stress decreases it and there may be no marked change in a rock's mobility. However, decrease of effective stress always destabilizes a rock; a particularly likely cause is a transient rise in fluid pressure, other parameters staying constant.

Change of depth inexorably changes a rock's behavior in significant ways at a slow rate; change of fluid pressure is a much more fast-acting control.

Change of fluid *amount,* i.e., change of porosity, is also a fast-acting agent, but tends to be self-terminating, because increase in the amount of fluid at a site is very commonly accompanied by increase of fracturing, and the fluid escapes.

The view emerges that rocks range from being rather frictionlike to being rather fluidlike. The composite intermediate behavior can be treated by numerical mechanics, Figure 7.10 being an indication of how one might begin (or of course a version with smooth curves could embody the same essential properties). The view that rocks mostly

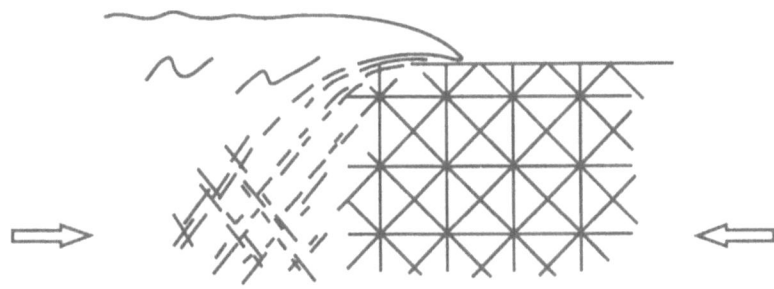

FIGURE 7.14

do not behave at either the frictional or the fluidlike extreme is symbolized by renouncing the sharp boundaries in Figure 5.5 and drawing Figure 7.14 instead.

It is an appropriate moment to recall the overall intention underlying this book. We seek models or representations of the behavior of rocks that are sufficiently simple to be handled numerically but yet sufficiently rich and versatile to give a semblance or shadow of real rock behavior. Figure 7.10 is thought to have roughly the right degree of complexity, if we bear in mind first the additional effect of temperature, but bear in mind second and perhaps more importantly the changes in texture a rock undergoes—how different it can be at the end of a deformation from what it was when deformation began, and how different it can be during those transient middle stages (never preserved in outcrops) when it is expanded and destabilized by passage of a pulse of fluid.

ANSWERS

Answer 7.1

See Figure 7.15.

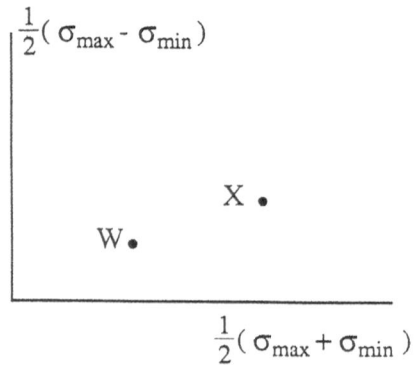

FIGURE 7.15

Answer 7.2

(a) See Figure 7.16: starting from a point representing σ_{min} at 30 MPa on the horizontal axis, a line sloping upward at 45° cuts the failure line at the needed crest-point and, from there, a line sloping down at 45° leads to σ_{max} at 210 MPa.

(b) For Mohr circles, the axes are for shear stress and normal stress, but for crest-points the axes are for $(1/2)(\sigma_{max} - \sigma_{min})$ and $(1/2)(\sigma_{max} + \sigma_{min})$.

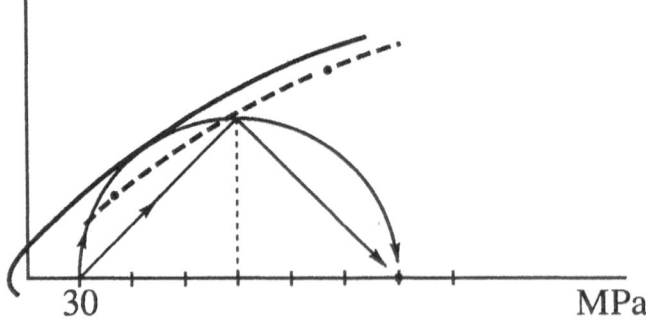

FIGURE 7.16

Answer 7.3

See Figure 7.17: the crest-points are at $(-7, 1)$ $(-4, 4)$ and $(-2, 6)$, on a line sloping upward at 45° from the point -8 MPa on the horizontal axis. Eventually this line is cut off by, or merges into, the line for shear failure.

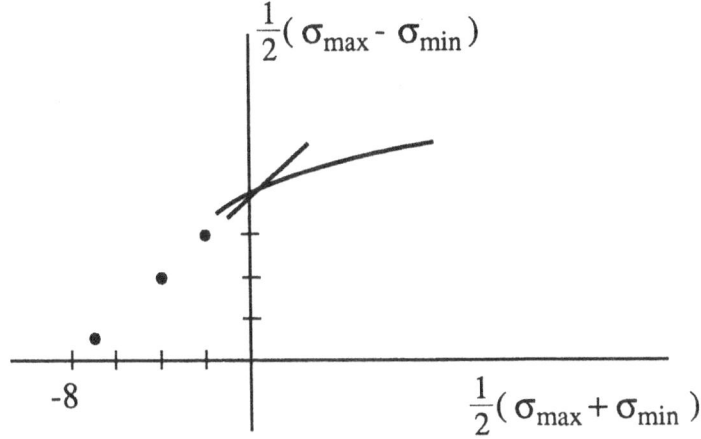

FIGURE 7.17

Answer 7.4

(a) (i) 10,000 per gtu or 10^{-10} per sec.
　　(ii) about 1000 per gtu or 10^{-11} per sec.

(b) (i) At 10 km, critical value of $\sigma_h - \sigma_v = 75$ MPa and $\sigma_v = 300$ MPa; therefore, $\sigma_h = 375$ MPa.
　　(ii) $\sigma_v - \sigma_h = 65$ MPa, therefore, $\sigma_h = 235$ MPa.

Answer 7.5

At 21 km on the horizontal axis in the section for P_{fluid} of 600 MPa—i.e., the front section of the diagram—the vertical distance up to the contour for strain rate 10 per gtu corresponds to about 16 MPa on the vertical axis.

At 21 km depth, if $\sigma_{\text{vert}} = 630$ MPa, $\sigma_{\text{horiz},NS}$ of 646 MPa and $\sigma_{\text{horiz},EW}$ of 614 MPa would give the specified strain rate of 10 per gtu.

Answer 7.6

When we go to shallower depth at fixed P_{fluid}, we go to a smaller effective stress $\sigma_{\text{total}} - P_{\text{fluid}}$. The rock is close to fracturing; in fact the effective minimum stress $\sigma_{EW} - P_{\text{fluid}}$ is dropping from 14 MPa down toward zero. Despite the large *total* stress, an effective compression that is already small and is diminishing allows cataclastic flow to speed up.

When we go to greater depth at fixed P_{fluid}, effective compression goes up and cataclastic flow is suppressed. But the rock is already at a temperature of 525°C, and doing deeper means getting hotter. For the dislocation-creep processes that now dominate the action, rise of temperature accelerates them more than rise of pressure slows them down.

Answer 7.7

The behavior is basically of the same temperature-sensitive type in both diagrams. It is a little more sluggish at lower fluid pressure; for example, at 23 km depth, 20 MPa of shear stress produces a strain rate of about 150 per gtu at the higher fluid pressure and 30 per gtu at the lower fluid pressure or higher effective compression.

Answer 7.8

Easy comparisons to make using the contours provided are at 16 km for pressure solution and 23 km for dislocation creep. At 16 km, doubling the stress difference from 40 MPa to 80 MPa just doubles the strain rate from 1 to 2 per gtu, whereas at 23 km, doubling the stress difference from 10 MPa to 20 MPa triples the strain rate from 10 to about 30 per gtu. At higher temperature the contrast is more marked: the contours suggest that by 25 km depth and 625°C, where about 3 MPa of stress difference gives strain rate 10 per gtu, doubling the stress difference could give a strain rate 10× larger.

Answer 7.9

(a) (i) $6000 - 2400 = 3600$ per gtu $= 1.1$ per 1000 years
    (ii) $9000 - 9600 = -600$ per gtu $= -0.2$ per 1000 years
(b) (i) $8000 - 2400 = 5600$ per gtu $= 1.7$ per 1000 years
    (ii) $12{,}000 - 9600 = 2400$ per gtu $= 0.7$ per 1000 years
          $36{,}000 - 19{,}200 = 16{,}800$ per gtu $= 5$ per 1000 years

(c) At either level of τ, we see that as the number of weaknesses increases from 2 to 8 per mm³, both rates of change increase: shear stress creates weaknesses faster *and* compressive stress heals them faster. The significant difference is that when $\tau = 18$ MPa, there are values of w for which healing can eliminate weaknesses faster than shear stress creates them; in fact, at $\tau = 18$ MPa, the number of weaknesses never reaches 7 per mm³ if it starts at some low value such as 2 per mm³—healing balances creation at a concentration just below 7 per mm³ and the population stabilizes at that value.

By contrast, if $\tau = 24$ MPa, the *least* value of dw/dt is more than 0.6 per 1000 years, (when w is about 9 per mm³); the higher shear stress creates weaknesses at a rate that is never matched by the healing process. It takes nearly 2000 years for w to climb from 9 per mm³ to 10 per mm³, but after that, the rate at which shear stress outstrips healing increases. Once w reaches 16 per mm³, it can be seen that the critical value of 20 per mm³ when the failure process completely takes over is less than 1000 years away.

Answer 7.10

(a) Average compressive stress at back end = 42 MPa; ratio of areas = 5; therefore, average shear stress at base = 8.4 MPa.

(b) Strain rate in sole = 200 per gtu;
    Viscosity = 8.4/200 = 0.042 gvu.

Answer 7.11

(a) 16.8 MPa
(b) Under 5 km of overburden, the vertical compressive stress on the slip surface is 140 to 160 MPa; thus an effective coefficient of 0.12 or less is needed.

(c) We need effective stress = 1/5 × total stress, and thus we need fluid pressure = 4/5 × total stress.

Answer 7.12

There is no standard answer; the question is left as an exercise for the reader. Some tentative thoughts follow in the main text, but the purpose of the book has been accomplished if readers are willing to attack Question 7.12 on their own.

Answer 7.13

This question is exactly on the mainstream of the chapter's development so, for a suggested answer, please return to the main text.

Appendix A: List of Abbreviations

cm	centimeter
ft	foot
g	$\begin{cases} \text{gram} \\ \text{acceleration due to gravity,} = \text{number of newtons exerted by 1 kg.} \end{cases}$
gtu	geological time unit
gvu	geological viscosity unit
in.	inch
int	intermediate
kb	kilobar
km	kilometer
lb	pound
l.h.	left-hand
$\log_n$	natural logarithm
lst	limestone
m	meter
max	maximum
min	$\begin{cases} \text{minimum} \\ \text{minute} \end{cases}$
mm	millimeter
μm	micrometer
MPa	megapascal
my	million years
N	newton
Pa	pascal
psi	pounds per square inch
r.h.	right-hand
sec	second
t	metric ton

tfd	total force down
tfu	total force up
v	versus
vol	volume
yr	year

Δ	Δx = a small but finite change in x or a difference between two x-values
μm	micrometer

Appendix B: Precision and Significant Figures

The general rule throughout the book is to show two significant figures. Rocks are so variable and their properties so poorly known that every calculation carries the thought "Let's try this as a possible value and see what the answer comes to." To write three significant figures would be to enter a dream world; one would have lost touch with the reality of the topics we are trying to discuss.

In some of the exercises several answers are calculated in sequence, later answers depending on values calculated earlier. Later answers then may contain round-off errors, or may look wrong. For example, successive answers may be given as

displacement rate	1.6 mm/yr
travel time	25 yr
total displacement	39 mm.

A reader who has figured the total displacement as 40 mm is asked not to worry. The important things to agree are (i) that the calculation can be done, so let's do it, and (ii) that the answer comes somewhere between 35 and 45 mm.

The underlying belief is that there are many opportunities for the reader to do original research in structural geology using the back of an envelope.

Appendix C: Weights as Units of Force

One kilogram is an informal but a very familiar unit of force. A mass of 1 kg at the earth's surface exerts a force of about 9.8 newtons; a newton is a wholly formal but slightly less everyday unit of force. Readers are encouraged to use "the weight of 1 kg" or 10 N, whichever they are more familiar with.

The most frequent reason to consider a rock's weight is to estimate the pressure or stress underneath it: to generate 1 MPa of vertical stress requires 30 or 40 m of rock overburden. It does not matter whether one thinks of this as 100 N/cm^2 or 100 tons/m^2 or 10 kg/cm^2.

An objective of this book is to make a link between geological structures and everyday experience, and for many people the link is through the kitchen: the nearest analog for a deformable rock is some kind of deformable food. Now a kilogram is a kitchen unit whereas a newton is not. Readers who work in newtons are asked to give particular attention to the opening of Chapter 3 and to maintaining links to daily experience. For example, how high can one stack unconfined ice-cream before it collapses under its own weight? It is not important whether one uses newtons or kilograms in working out an answer (or a purely experimental approach); the important thing is to realize that behavior of rocks is *familiar behavior*—it is *like* behavior we have seen in other materials. To use a kilogram as a unit of force is somewhat favored for the sake of familiarity.

Appendix D: Experimental Work on Rock Deformation

In the main text, rock properties are usually guessed at or assumed, and little reference is made to measured values. But the objective of the book is that when we see a rock structure in outcrop, we should be able to form some conjecture about its process of formation. If this is the objective, should not *measuring* the properties of the rock we see play a part? Should not measured properties guide us, and in fact limit us, in the conjectures we form? The purpose of this appendix is to consider the role of measured values. The conclusion is that we certainly need them but that we also need carefully considered guesses and trial values.

The central difficulty is the time scale. There are other difficulties:

1. We need to recreate the temperature, total pressure, pore-fluid pressure and grain texture that existed at the time the rock deformed.
2. Recreating a pore fluid like the one that occurred in nature tends to do the laboratory equipment harm through corrosion; protections and vigilance are needed.
3. We don't know what grain texture and what pore-fluid chemistry we need to recreate. It is safe to assume that the texture we see today—in particular the microfractures we see today—are not the texture and microfractures that were present during deformation.

The third point brings us back to the central difficulty, the time scale. A typical program of experiments occupies less than 10 years and fewer than 10 deformation machines. A realistic question that faces an investigator is, "What shall I attempt with this machine this coming year—one or two experiments lasting for months or a greater number lasting a few days each?" For any of the variables such as composition or texture, to discover the effect of the variable one needs a *series* of experiments; hence the investigator's desire to cover a range of magnitudes for some variable conflicts with the desire to spend four months on each experiment. There are not many examples of a systematic series of tests designed to explore the effect of a single variable where each test in the series lasts for months.

A consequence is that rocks in experiments are deformed more rapidly than in most natural situations. Suppose a rock is strained by 10 percent in four months: the strain rate is 10^{-8} per sec or 10^6 per gtu. This is faster than most geological deformations by at least 10,000 times. The difference this might make can be illustrated by thinking of a caterpillar-track vehicle in rough country proceeding over obstacles at 1 mile an hour. If there were a means of making the same vehicle cross the same terrain at 10,000 miles per hour, obviously the *nature of the interaction* between vehicle and ground surface would be very different.

Of course it is not only experiments that can be misleading; analogies can be misleading too. To consider more carefully the difference that change of time-scale might make, we should give separate attention to fracture behavior and flow.

Fracture Behavior   Two points emerge from the large number of fracture experiments that have been performed.

1. For many types of rocks, if the minimum compressive stress is of the same order as overburden stresses in the crust (30 to 1000 MPa) the failure envelope slopes at about 35°; i.e., the tangent of the angle of slope is about 0.7.
2. The sample begins to make cracking noises at about half the stress difference at which it later fails.

The data on which the first generalization is based are gained in experiments lasting a few minutes or up to a few hours. Let us consider: suppose a sample failed in a two-hour test at a stress difference of 120 MPa and we subjected a similar sample to a stress difference of 90 MPa. Sensitive microphones pick up signals from little cracking events inside the sample. If we maintain the stress difference for 1000 years, what is the likelihood of the rock fully coming apart during that time?

The result of an experiment lasting 1000 years can only be guessed at, but the sample might very well fail. Thus the experimental result is an *upper limit* to the property sought. If rock behavior in the laboratory is described by an envelope whose slope is 0.7 and rocks in nature fail at three-fourths of the laboratory failure stress, rock behavior in nature would be described by an envelope whose slope is 0.5. We do not know *how large* a correction of this type needs to be made, but the noise of incipient cracking suggests that some such correction would be realistic.

A correction of the type just suggested would be significant but is still only a change of 20 or 30 percent; it is not a change by an order of magnitude, and we know in which *direction* to go; the value of 0.7 seems robust as an upper limit—natural values can only be *less*. With flow behavior, it will be suggested first that the uncertainties are larger, and second that we are not sure whether an experiment overestimates or underestimates a rock's natural response.

Flow Behavior   Essentially the same problem arises again: we wish to perform experiments at strain rates of 10^{-5} to 10^{-8} per second (10^9 to 10^6 per gtu) and deduce from them behavior at rates at least a million times slower. The possibility of the experiments underestimating a natural rock's response-rate is discussed first, and the possibility of overestimating the response rate is discussed later.

The possibility of underestimating a rock's response arises because, when stressed, a rock may respond in more than one way. For example, some grains may slip past each other like marbles in a bag while at the same time, other grains change shape by pressure solution. Each of the processes will run faster if the driving stress difference increases, but one process may be affected more than the other. That is, a rock is likely to respond by two mechanisms (or more), and one mechanism will be more stress-sensitive than the other. In such a case, experimental results may be dominated by one mechanism while slower natural deformation of the same rock is dominated by the other mechanism. For a numerical illustration, see Figure D1.

In the situation illustrated, it is supposed that one response-rate in the rock is proportional to the driving stress difference itself, whereas the other response-rate is proportional to (stress difference)^3. At high stress the total response is indistinguishable from just the second effect, whereas at lower stress the total response is indistinguishable from just the first effect. Experiments conducted in the high-stress range would not provide information about the process that dominates the rock's behavior at lower stresses.

A closely similar situation exists where the variable is temperature. Again, for the sake of a detectable strain rate, an investigator might work at a high temperature and then try to extrapolate the trend of the data to natural strain rates at lower temperatures. Again there is the possibility that a mechanism that plays an insignificant role at high strain rates plays a dominant role at lower strain rates. The experiments would then reveal what we don't need and fail to reveal what we do need.

If the preceding difficulties were the only obstacle, the situation would perhaps be manageable, because the experimental results would provide at least a limit. As noted, the only consequence of the effects discussed is to *under*estimate the rock's mobility. One could be confident that a rock's mobility was *at least* as great as the experiments suggest. But unfortunately there is another effect that is hard to explore by which the experiments might overestimate a natural rock's mobility, as follows.

The point to be recalled was made in Chapter 7 and underlies Figure 7.5—a rock's mobility is affected by competing influences: compressive stress promotes healing of weaknesses, whereas shear stress tears a rock apart. A feature of experimental results is

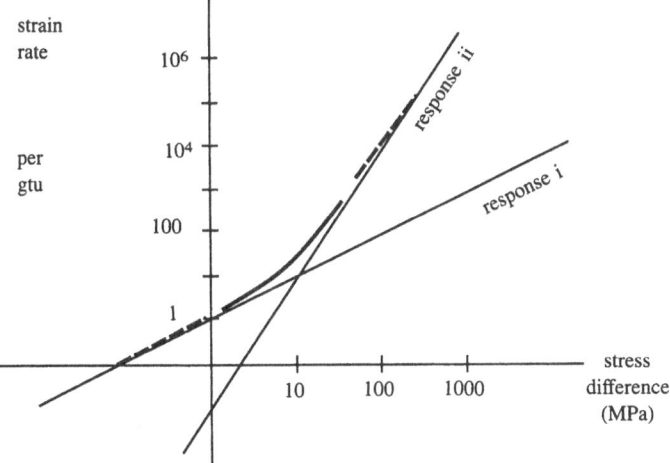

total response = sum of (i) + (ii) → heavy line

FIGURE **D1**

that a rock's flow behavior seems insensitive to the amount of compressive stress, but if experiments could be run more slowly, this apparent insensitivity might disappear. Suppose that before an experiment was performed, a rock could be kept undisturbed at 500°C and 500 MPa for a thousand years; then a small, slow deformation imposed, and then another thousand years of healing and another small deformation: the behavior that would result simply cannot be discovered. I suppose a rock deformed in that manner would be less mobile than a similar rock deformed without the healing intervals, but would the difference be large or small? There is no way to tell. (It is possible that the question can be illuminated by considering *cratons*. These cores of continents—the Canadian shield, Baltic shield, etc.—seem exceptionally durable. Perhaps their special quality comes in part from not having been disturbed for a long time, from thixotropy on a geological time scale.)

Clearly we are now far into the realm of speculation and should return to the main text. The reason for returning is that experiments cannot tell us all we want to know; it is necessary to *combine* indications from experiments with indications such as the text contains: some sets of trial values and interrelations seem to hang together in a coherent story, while other sets don't work so well. To put trial numbers into some mechanical theory and see how the arithmetic works out is one means of learning about rocks. The method in the main text, of trial values and speculations, and the method of experiments need each other. In combination, they will increase the frequency of occasions when we can look at an outcrop and feel, "Yes, I have a possible idea why that came out the way it did."

Appendix E: Books on Related Material

Books in the following list have been selected because they deal with topics from the main text at a comparable level. They can be read concurrently or following the main text, and will lead the student to more advanced books and the journal literature.

Beside omitting more advanced books, the list omits books covered by the general term *geodynamics,* which treat larger-scale features of the earth's structure. Our main concern here is what can be seen in outcrops or, at most, on a mountainside.

Stress and Strain by W. D. Means (Springer-Verlag, 1976) overlaps mainly with Chapters 2, 3, and 4. It is exceptionally thorough and clear, with many exercises with solutions provided. Matrix methods are fully explained; readers wishing to supplement the summaries on pages 71 and 98 should turn to this book.

Elasticity, Fracture and Flow by J. C. Jaeger (3rd edition, Methuen, 1971) treats the topics almost wholly in algebraic form. In the present book, a topic is introduced by a numerical example and an algebraic statement comes later, whereas Jaeger begins with the algebraic form and pursues its ramifications. In large part, the purpose of the present book is to enable readers to use Jaeger's book with pleasure and profit; it is uniquely compact and well organized, a valuable reference and resource.

Fundamentals of Rock Mechanics by J. C. Jaeger and N. G. W. Cook (3rd edition, Chapman and Hall, 1979) presents the same fundamental equations in a less tightly knit and more discursive form.

Structural Geology by M. K. Hubbert (Hafner Publishing, 1972) is a collection of separate papers rather than a monograph or textbook. Thus only selected topics are covered, but the treatment is very thorough and begins from the same basic concepts of force and stress as the present text. It is recommended for reading particularly as an extension of Chapter 6.

Mechanics of Tectonic Faulting by G. Mandl (Elsevier, 1988) covers roughly the same topics as Hubbert's book. A person who has worked through

Stress and Strain by Means and the simpler parts of Hubbert's book will find that Mandl goes more deeply into certain aspects. (The arrangement of the book is odd, in that the basic concepts are covered in part 2, after they have been used in the extensive and valuable section on applications in part 1.)

The Techniques of Modern Structural Geology by J. G. Ramsay and M. I. Huber (Academic Press, vol. 1, 1983; vol. 2 1987) treats strain (vol. 1) and folds and fractures (vol. 2). Volume 1 expands Chapter 2 of this book in a manner that is both very thorough and very rich in links to real rocks; the illustrations impel the user to delve into the text. Volume 2 has less overlap with the present text but presents a wealth of instances where quantitative mechanics could be applied in an exploratory way. (The fractures discussed are on small scale; there is little overlap with the book by Mandl either in topic or approach.)

Principles of Rock Deformation by A. Nicolas (D. Reidel Publishing, 1987) provides a briefer coverage of the topics that Ramsay and Huber treat in depth. The book helps the reader to see what processes seem to have occurred in rocks, with little attention to questions about the stress field that drove them or the rock's response as regards overall strain rate or effective rheology.

Analysis of Geological Structures by N. J. Price and J. W. Cosgrove (Cambridge University Press, 1990) is an excellent compendium, which gives mechanics-based equations for many geological structures; readers seeking a survey of instances where ideas from the present book have been applied will find many examples here. But the authors generally do not derive the equations quoted; one cannot use this book to *continue learning* how to apply the fundamental ideas. Price and Cosgrove assume that people with that ambition will manage elsewhere.

This list would be incomplete without mention of *Introduction à l'Etude Mécanique des Déformations de l'Ecorce Terrestre* by J. Goguel (2nd edition, Imprimerie Nationale, Paris, 1948). This book is hard to find and has never been published in translation, to the impoverishment of geological science. Almost 50 years ago, Goguel created this comprehensive and incisive masterwork; there is something wrong with the practice of geological education that such a source-book has been so little used. Readers who can find a copy are encouraged to grapple with it and enjoy it, without necessarily proceeding from page 1 forward.

Index

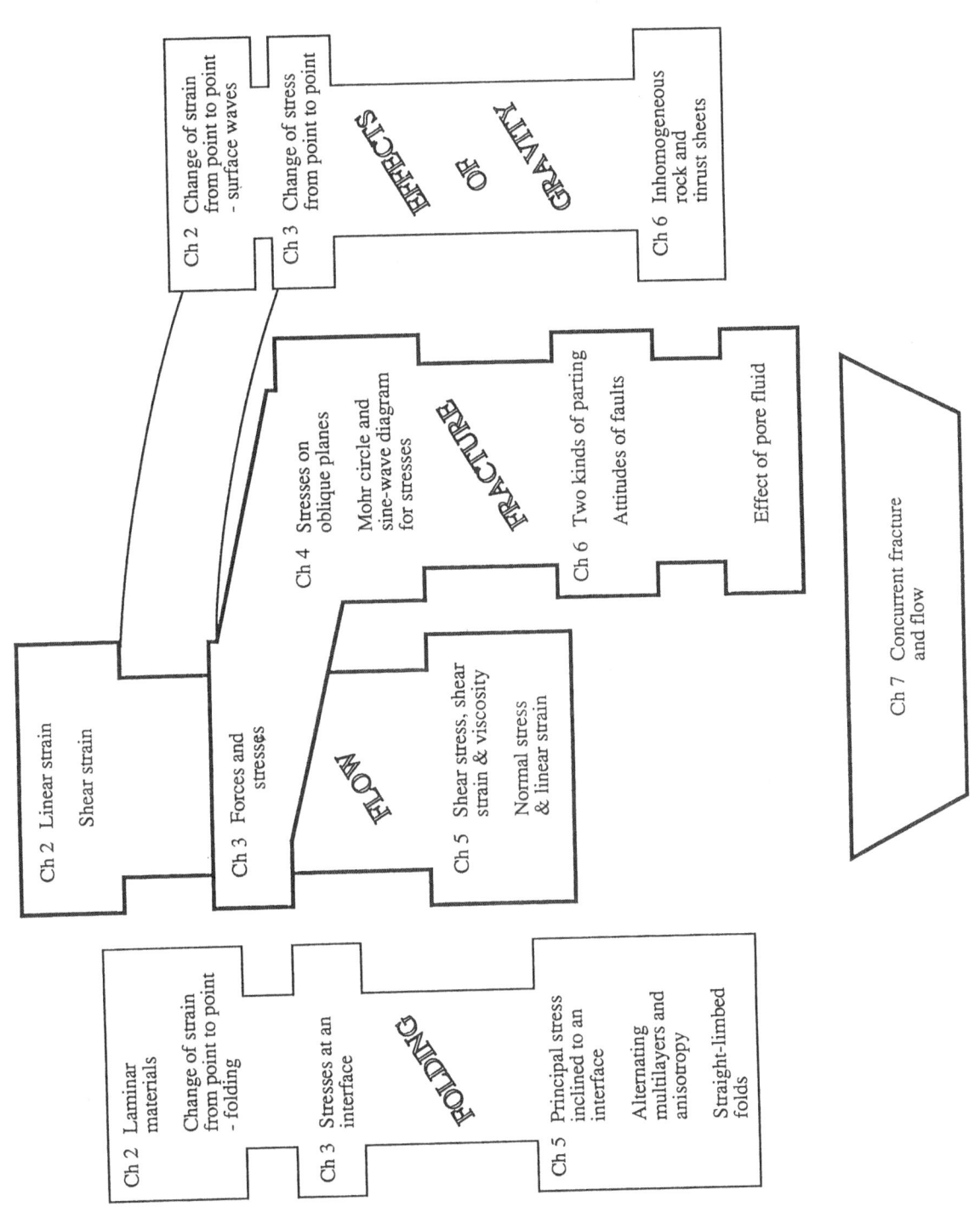

EFFECTS OF GRAVITY

Ch 2 Change of strain from point to point - surface waves

Ch 3 Change of stress from point to point

Ch 6 Inhomogeneous rock and thrust sheets

FRACTURE

Ch 4 Stresses on oblique planes

Mohr circle and sine-wave diagram for stresses

Ch 6 Two kinds of parting

Attitudes of faults

Effect of pore fluid

Ch 7 Concurrent fracture and flow

FLOW

Ch 2 Linear strain

Shear strain

Ch 3 Forces and stresses

Ch 5 Shear stress, shear strain & viscosity

Normal stress & linear strain

FOLDING

Ch 2 Laminar materials

Change of strain from point to point - folding

Ch 3 Stresses at an interface

Ch 5 Principal stress inclined to an interface

Alternating multilayers and anisotropy

Straight-limbed folds

Units

Mass	g		
	kg	lb	454 g
	t metric ton 1000 kg	ton 2000 lb $\cong$ 1 metric ton	

Force   newton   1 newton accelerates 1 kg at 1 m/sec²
dyne   1 dyne accelerates 1 g at 1 cm/sec²

(lb, g, and kg as forces: the weights of a pound, gram, or kilogram are convenient force units, though informal and somewhat variable according to the local gravity field.)

Acceleration due to gravity   9.8 m/sec²   or   981 cm/sec²
Thus the weight of 1 kg accelerates 1 kg at 9.8 m/sec², so that the weight of 1 kg = 9.8 newtons and 1 newton $\cong$ the weight of 100 g.

Pressure   pascal   1 newton/m²
KPa

MPa $1 \text{ bar} = \frac{1}{10} \text{ MPa}$

                                 $1 \text{ kb} = 100 \text{ MPa}$

GPa

1 MPa = 10 bar = 145 psi = 10 kg/cm² = 100 t/m²

10 m of water $\cong$ 3 or 4 m of rock $\cong$ 76 cm of Hg $\rightarrow$ 1 bar
  or   or   or
30 ft   10 ft   30 in

                    30 or 40 m of rock $\rightarrow$ 1 MPa
                    1 km of rock $\rightarrow$ ~30 MPa

Energy
(1) force $\times$ distance   1 newton-meter = 1 joule
(2) pressure $\times$ volume   1 pascal-meter³ = 1 joule
(3) (heat capacity) $\times$ (change in temperature). To heat 1 g of water through 1°C requires 4.2 joules.

Time
1 gtu = 10^{14} sec $\cong$ $3\frac{1}{3}$ my
1 my $\cong$ $3 \cdot 10^{13}$ sec
1 year $\cong$ $3 \cdot 10^7$ sec   (more exactly, $3.156 \cdot 10^7$ sec)

Viscosity
1 gvu = 1 MPa-gtu = 10^{14} MPa-sec
       = 10^{15} bar-sec = 10^{20} Pa-sec = 10^{21} poise

1 MPa-sec = 10 bar-sec = 10^6 Pa-sec = 10^7 poise
          1 bar-sec = 10^5 Pa-sec = 10^6 poise